12th International Conference on Building Defects (Building Defects 2020)

Held online due to COVID-19

IOP Conference Series: Materials Science and Engineering
Volume 1066

Ceske Budejovice, Czech Republic
26 - 27 November 2020

ISBN: 978-1-7138-3342-0
ISSN: 1757-8981

Printed from e-media with permission by:

Curran Associates, Inc.
57 Morehouse Lane
Red Hook, NY 12571

Some format issues inherent in the e-media version may also appear in this print version.

Institute of Physics
Dirac House, Temple Back
Bristol BS1 6BE UK

Phone: 44 1 17 929 7481
Fax: 44 1 17 920 0979

techtracking@iop.org

Additional copies of this publication are available from:

Curran Associates, Inc.
57 Morehouse Lane
Red Hook, NY 12571 USA
Phone: 845-758-0400
Fax: 845-758-2633
Email: curran@proceedings.com
Web: www.proceedings.com

TABLE OF CONTENTS

IOP Conf. Series: Materials Science and Engineering 1066 (2021) 011001 doi:10.1088/1757-899X/1066/1/011001

Building Defects 2020

Ingrid Juhásová Šenitková [1], Michal Kraus [1]

[1] Institute of Technology and Business in České Budějovice, Department of Civil Engineering, Okružní 517/10, 370 01 České Budějovice, Czech Republic

Infobuildingdefects@gmail.com

Preface

We are pleased to welcome you to the conference proceedings of the 12th International Scientific Conference on Building Defects (Building Defects 2020) held on 26th – 27th November 2020 at the Institute of Technology and Business (ITB) in České Budějovice, Czech Republic. The conference is organized on behalf of His Magnificence prof. Marek Vochozka, MSc., MBA., Ph.D., rector of the Institute of Technology and Business in České Budějovice and under the honorary patronage of the Czech Chamber of Certified Engineers and Technicians (CKAIT).

The current conference is the twelfth continuous of Building Defects conferences, which solves especially the causes of defects during the building processes and afterward defects of the building themselves as well as the defect of building structures. The conference is the premier chance for the presentation of new advances and research results in the fields of civil engineering. The main aim of the conference is to present the scientific research results and its practical solutions in the field of civil engineering, taking into account the principles of sustainable development as well as to present contemporary civil engineering issues and future development.

It is our pleasure to announce the participation of expert presenters from various universities and research institutes in this two-day event. A lot of thanks are given to all authors of submitted papers for their participation. They contributed a great deal of effort and creativity to produce this work. We are happy they choose the 12th Anniversary International Scientific Conference on Building Defects to present their excellent work. Thanks also to all editors and reviewers, who donated substantial time from their busy schedules to carefully read and evaluate the submitted scientific papers. The conference proceedings contain 19 papers. Each contributed paper was rigorously reviewed by two reviewers who were drawn mainly from the International Scientific Committee.

The spread of the coronavirus COVID-19 made adjustments to our life in this year, including the organization of our conference. Due to the COVID-19 pandemic situation all around the globe, we have organized this conference into a virtual conference and the method is a virtual presentation for all international and local participants. Participants could take part in the conference in video conference or of poster presentation remotely just like in the case of traditional face-to-face format. Everyone had an opportunity to watch the stream presentation or poster presentation and ask questions in chat, and speakers were able to give their answers and comments.

At last, but not at least, we would like to appreciate to all supporting organizations for their support and encouragement and got making the event a success. In addition, I would like to thank also to Organizing Committee for their great preparation work. We are looking forward to the next 13th International Conference Building Defects 2021 that will be held next year in November. We hope that it will be interesting and enjoyable at least as all of its twelfth predecessors.

prof. Ingrid Juhásová Šenitková, MSc., CSc.
Conference Guarantee
ITB in České Budějovice

International Scientific Committee

- prof. Marek Vochozka, MSc., MBA., Ph.D.

  Rector of ITB in České Budějovice, Czech Republic

- prof. Ingrid Juhásová Šenitková, MSc., CSc.

  Conference Chair (main editor), ITB in České Budějovice, Czech Republic

- prof. Haghighat Fariborz, MSc. Ph.D.

  Concordia University, Canada

- prof. Jan Váchal, MSc., CSc.

  ITB in České Budějovice, Czech Republic

- prof. František Kulhánek, MSc., CSc.

  Czech Technical University in Prague, Czech Republic

- prof. Pavol Juhás, MSc., DrSc.

  ITB in České Budějovice, Czech Republic

- prof. Jan Mareček, MSc., DrSc., dr. h. c.

  Mendel University in Brno

- prof. Jozef Oláh, MSc.

  STU in Bratislava, Slovakia

- prof. Marián Flimel, MSc. Ph.D.

  TU in Košice, Slovakia

- prof. Josef Jurman, MSc., CSc.,

  VŠB - Technical University of Ostrava, Czech Republic

- prof. Jindřich Melcher, MSc., DrSc.,

  Brno University of Technology, Czech Republic

- prof. Šárka Šilarová, MSc., CSc.,

  Czech Technical University in Prague, Czech Republic

IOP Conf. Series: Materials Science and Engineering 1066 (2021) 011001 doi:10.1088/1757-899X/1066/1/011001

Organizing Committee

- Jan Plachý, MSc., Ph.D.

  Conference Program Director, ITB in České Budějovice, Czech Republic

- Michal Kraus, MSc., Ph.D.

  ITB in České Budějovice, Czech Republic

- Vojtěch Stehel, MSc., MBA, Ph.D.

  ITB in České Budějovice, Czech Republic

- Robert Šinkner, MSc., MBA

  TKP geo s.r.o., Czech Republic

- Petra Machová, MSc. et MSc.

  ITB in České Budějovice, Czech Republic

IOP Conf. Series: Materials Science and Engineering 1066 (2021) 011001 doi:10.1088/1757-899X/1066/1/011001

Sponsors

- Czech Chamber of Certified Engineers and Technicians (CKAIT)
  Conference general partner
- Journal Střechy-Fasády-Izolace
  General medial partner
- Internet portal Izolace.cz
  Medial partner
- Internet portal iMaterialy.cz
  Medial partner
- Journal Build Info
  Medial partner

Peer review declaration

All papers published in this volume of *IOP Conference Series: Materials Science and Engineering* have been peer reviewed through processes administered by the Editors. Reviews were conducted by expert referees to the professional and scientific standards expected of a proceedings journal published by IOP Publishing.

- **Type of peer review:** Double-open
- **Conference submission management system:** Internal University System
- **Number of submissions received:** 29
- **Number of submissions sent for review:** 29
- **Number of submissions accepted**: 21
- **Acceptance Rate (Number of Submissions Accepted / Number of Submissions Received X 100):** 72.41%
- **Average number of reviews per paper:** 2
- **Total number of reviewers involved:** 12
- **Any additional info on review process:**

  The review process consists of two phases. First, each of the submitted abstracts is reviewed by the guarantee of the conference or an authorized expert. If the abstract meets the criteria and is in accordance with the focus of the conference, a full paper is requested.

  Each of the submitted full papers is reviewed by two professional experts in the related subject area. A report of the reviewing result will come out by each expert. The review report included ratings on the following points: Is the manuscript of high scientific quality? | Does the paper contain enough new results? | Is the research design appropriate? | Is the manuscript free from errors? | Is the paper scientifically sound and not misleading? | Is the paper clearly written, concise and understandable? | Is the title appropriate? | Are the methods adequately described? | Are the results clearly presented? | Are the conclusions supported by the results? | Are the references to related work adequate? | Should the English writing of the manuscript be improved? | Are the figures clear? | Are the tables clear?. Each review report contains a final recommendation for the editor (Publish as is | Publish after optional minor revision | Publish after mandatory minor / major revision | reject). The papers which require revision are sent back to the authors to incorporate the corrections and changes as suggested by the reviewers. The revised papers are again reviewed by the editor and recommended for publication. The accepted papers have only been called for the presentation and recommended for publication in IOP Conference Proceedings.

IOP Conf. Series: Materials Science and Engineering 1066 (2021) 011002 doi:10.1088/1757-899X/1066/1/011002

- **Contact person for queries:** Michal Kraus, 19514@mail.vstecb.cz

Building Defects 2020 IOP Publishing
IOP Conf. Series: Materials Science and Engineering 1066 (2021) 012001 doi:10.1088/1757-899X/1066/1/012001

Structural and Building Engineering – an Introduction

Ingrid Juhásová Šenitková,[1] Pavol Juhás[1]

[1] Institute of Technology and Business in České Budějovice, Department of Civil Engineering, Okružní 517/10, 370 01 České Budějovice, Czech Republic

14667@mail.vstecb.cz

Abstract. Construction is a reflection of the current state of society. The construction is created as a result of a creative process including preparation, all stages of design and implementation. Current trends of modern building structures with a large span are characterized by the introduction of new types of load-bearing structural systems with excellent ecological, economic and aesthetic properties. The paper aim is to introduce the interaction and context in design and create a reliable economic and environmental friendly structure that will serve the required purposes.

1. Introduction

There is often a parallel between a prosperous society and a prosperous construction industry, or in other terms, where there is all-round economic development, construction is also developing, and vice versa, where the economy is stagnant, roads, bridges and buildings are not designed and built. Thus, construction is a reflection of the current state of society. It intervenes in all its development areas associated with capital expenditures, respectively with investment.

There is perhaps no human activity with which we would come into everyday contact as often and multilaterally as building. According to the purpose, we distinguish between buildings, resp. civil, engineering, transport, industrial, agricultural works, etc. Depending on the type of material used, we are talking about masonry, wooden, concrete, steel, plastic etc. Construction generally deals with building structures and works, preparation, design and implementation of buildings, building materials, building economics, etc. With the gradual development of architectural art and its knowledge, construction becomes a testimony to the development of society. The construction is created as a result of a creative process including preparation, all stages of design and implementation. It is necessary to realize that this process has its historical development and is historically connected with the period; it has its developmental connections and regularities. These are a source of new inspiration and knowledge. For general orientation in the field of building engineering, it is necessary to know the history of construction and architecture, development of materials, development of structures, systems and methods of construction, styles and their characteristic architectural and expressive means, aesthetics, development of functionality of buildings and their layout - operational solutions, technologies, as well as development of technical equipment of buildings. Through the synthesis of existing knowledge and their thorough analysis, we can arrive at a new quality, new shapes and forms, better construction systems, etc. There is another important reason why it is necessary for the current practical activity to know and study the historical development of buildings.

This reason is that historic buildings are part of today's architecture, which we functionally use, maintain, restore, reconstruct, etc. As a country cares about its historical cultural heritage, the maturity of its nation is measured.

Many new scientific and technical problems arise because technical works have to operate in increasingly demanding conditions, with ever-increasing sizes and loads. The growing quantity of relevant quantities is growing into a new quality, which requires new theoretical and experimental approaches, more complex models, new methods and algorithms in design and assessment.

2. Conceptual designs of the structures

Let's ask ourselves when the first conceptual design of a building structure was created. Today, it is generally accepted that Poleni, with his expertise in St. Peter's Dome in Rome, in 1748, carried out the first formal analysis of the stability of the structure, and thus a professional structural engineer was born. At the end of the 19th century, industrial, exhibition, transport and other buildings were covered with large-span three-joint arches [1].

The beginning of conceptual design of the structure can be characterized by the construction of the Palace of Machines of the World's Fair in Paris in 1889, which integrated new technological and scientific knowledge. The innovative construction, which summarized everything that was built before, had two principled original peculiarities. The first was the use of a new material - steel, which has excellent properties (high tensile strength) compared to the cast iron used until then. The second was the application of the newly discovered Winkler's analytical method for solving truss bar arches from 1868. It was also understood that the appropriate situation of the joints can handle the problem of additional forces from thermal effects and settling. The engineer became a conceptual designer. He was no longer limited in analyzing and verifying his design and was able to direct the flow of forces by appropriate placement of structural elements and details. He could imagine the behavior of the proposed structures and discover their new corresponding shapes.

The period of admirable masters - builders who built without knowledge of the theory of structures, who had a "feeling" but no knowledge of the rules of the design gradually ended and came the period of engineers - conceptual designers armed with new discoveries, ideas and tools.

The development of structures and technologies for the construction of buildings is part of the development of science and technology and thus the social development of mankind. It is therefore directly affected by their level in a given period. Until the beginning of the 20th century, the traditional material base dominated (buildings made of stone and brick and others). Realizations of steel building structures were unique in the 19th century. The decisive milestone in the development was the discovery of reinforced concrete and its use since the end of the 19th century. Reinforced concrete has gradually become the most widespread structural material for load-bearing structures of multi-storey buildings [1].

It is interesting to follow the development of engineering thinking and technical possibilities of implementation of very progressive and interesting construction solutions for the time being. Often limiting factors of engineering thinking and slick designs developed at the time were the technological possibilities and existing structural elements and materials. The precursors of creative ideas have always surpassed its time and its real possibilities. The power of the spirit often preceded the material conditions.

A beautiful example is the work of the Czech engineer B. Schnirch, who was the first in the world to be granted a patent for fiber roof structures in 1826. He built several such roofs on the territory of the former Czechoslovakia. In 1824, the engineer Schnirch designed the load-bearing system of the roof of the theater building in Strážnice, with floor plan dimensions of 30 x 76 m, which he described in the article "On wrought iron trusses, their lightness, low conversion cost and usability". The supporting system consisted of chains designed under the ridge and on all corners, to which weaker chains were connected instead of traditional wooden structures. Later, in 1964, K. Tang, Y.Tsuboi, and M. Kawaguchi designed a suspended roof structure for the Tokyo Olympics. The supporting system consists of an impressive rope structure with floor plan dimensions of 120 x 214 m. Here, however, modern technologies and new technical elements such as high-strength ropes and special dampers have already been used to eliminate the adverse dynamic effects of oscillations.

The ideas were thus similar, resp. the same, but the method of implementation was dictated by time and its technological possibilities, significantly supplemented by the progress made in the field of science and technology. Intuition and empire have grown into a science-based theory. Computational resources have been transformed into computer technology, which has become a powerful tool in the hands of civil engineers. We can find many such examples; let´s call them "analogies in thinking with the time lag of perfection of realization", in the analysis of technology and construction history. It is important for us that at present the time interval between the progressive idea of an engineer and its implementation, thanks to exponentially developing technological possibilities and scientific discovery, are rapidly shortening.

Current trends of modern building structures with a large span are characterized by the introduction of new types of load-bearing structural systems with excellent environmental, economic and aesthetic properties, with very flexible use for variable spatial shapes and different floor plans. By suitable alternation of various materials (steel, concrete, wood, composite materials based on glass fibres and plastics, high-strength fibres, etc.) and through rational positioning of elements or subsystems with required properties in global construction, we obtain highly efficient combined - hybrid systems. We can also find their wide application in building construction. When we design modern intelligent structures the multidisciplinary approach is required. Such structures also include mobile systems. Example of this is a mobile roof structure. It is type of structure in which part or all of the roof structure can move resp. relocate so that the object can be used for both short periods of the roof, when it is fixed in the open or closed state.

3. Building materials

A number of materials are used to create the construction work, from its birth in the design studio, through implementation, to protection from the surrounding environment, which are to ensure trouble-free function. For centuries, classic natural materials based on inorganic and organic substances, such as stone, sand, wood and the like, have been used. The principal artificial building material was and still is a brick that has changed its form, but the content remains the same. Cast iron and later steel began to be used in the 19th century, and the first half of the last century was marked by the development of concrete, reinforced concrete and later prestressed concrete. Recently, a large number of new materials have emerged, which are gradually penetrating the construction industry. These are concretes with special properties, insulating materials, plastics, composite materials and substances combined from classic materials with various chemical additives. At the same time, a number of new types of structural elements and structures appear, such as thin-walled and sandwich structural elements, the already mentioned hybrid systems, etc. All this requires increased demands on the quality of the materials used and the need for comprehensive knowledge of their behaviour under different conditions. To do this, it is necessary to control their physical and chemical properties and the laws that characterize them. The correct and optimal choice of necessary materials and products requires knowing their technical parameters, which depend on the origin of the material used, the technology of its production and possible changes in certain characteristics and properties due to the environment during the life of the construction [2].

At present, the period of significant growth in the consumption of materials is behind us and a clear trend is being used to use less material. The tendencies are towards recycling and reuse of materials, replacement with inert materials and the use of innovative modern construction systems that consume less of materials. The reason is not only the global saving of materials, but also the fact that materials will be more expensive because there will not be enough increasingly cost-effective energy. The energy intensity of processes and products is becoming an important factor in the material industry, which is characterized by the interactions between raw materials, energy and environment.

Nowadays, the term "sustainable development" is often used. It is a development that meets the needs of the present without compromising the ability of future generations to meet their own needs, so as not to devastate the diversity of nature and preserve the natural functions of ecosystems. In connection with construction, we talk about the creation of a healthy exposed environment and responsible management of it, based on the principles of efficient use of resources and environmental principles. Thus, the rational use of material resources, as well as the materials themselves, is closely linked to the problem of energy and environmental problems. The recognition of local and global environmental problems have stimulated considerable process which has been made during the past years toward a more complete understanding of design and construction requirements for buildings. The building design and construction toward sustainable practices were slowly shifted by awareness of environmental problems. Many environmental-friendly design and construction principles are widely accepted, but designers and constructors often fail to recommended practices or their clients fail to adopt state-of-the-art practices. Reasons for these failures include preference for and adherence to traditional practices, lack of expertise or short-term financial considerations. Meanwhile, awareness of environmental problems has produced an emerging shift toward purportedly green building practices that presume to include buildings that are healthy for their occupants [3].

The main purpose of buildings is to create a climate more suitable for persons and processes than outdoor climate. However in buildings and shelters it is not only the thermal climate that is changed. The climate shell also stops the free air movement. The dilution of pollutants from close to man pollutants sources is diminished. The environment within building is always more polluted from indoor sources as building materials and indoor activities than outdoor air. This was and is the basis of the need for ventilation and for discussions on indoor air quality. Consequently, the main aim of building ventilation is to create the indoor air quality more suitable for persons and processes than what naturally occurs in the unventilated building, and to reintroduce the positive effect of being exposed to the wind, to dilute and remove the pollutants that man himself, his activities, and the indoor surrounding produce.

Sustainable buildings design depends significantly on the used criteria for the indoor environment and building design and operation. Indoor environment also affects health, productivity and comfort of the occupants. Recent studies have shown that cost of the deteriorated indoor environment for the society, employer and building owner are often higher than the cost of energy used in the same buildings. An energy declaration without a declaration related to the indoor environment makes no sense. There is therefore a need for specifying criteria for the indoor environment for design, energy calculations, performance and operation [4].

When we realize that about half of the price of buildings is the price of materials, it is clear that knowledge of the nature and properties of materials is very important for construction. For various structures and their parts, construction materials with different properties are needed, but common and most important are the properties ensuring quality and durability during the specified period of operation of the object.

The future is mainly focused on the purposeful preparation of new, composite materials - composites, materials with new properties and significantly more efficient use of matter and energy. A composite is any material system which consists of several phases (at least two), of which at least one is solid, with a macroscopically recognizable interface between the phases and which achieves properties which cannot be achieved by any component alone or by simple summation. Composite materials are combined from all major classes, e.g. metals, ceramics and other inorganic materials, polymers, into the final composite structure. The engineer must see in these materials a specific system instead of a simple mass with certain properties.

The main prerequisite for a significant increase in the efficiency of the structure is the combination of material design and structural design. Traditional methods of calculating the load-bearing capacity of a structure are changing.

Development is increasingly focused on the intelligent materials, which replace simple traditional materials with substances with higher material and energy efficiency. For example, polymer or metal composites reinforced with different fibres provide several times higher strength and stiffness than the best alloys, even in cases of high temperatures. The insulation board made of integral polyurethane foam insulates 25 times better than a brick wall with the same thickness, with a weight of 750 times less.

The using of material composites (carbon-fibre-reinforced plastics, ceramic-fibre-reinforced metals, fibre-reinforced concrete, etc.) and development of composites for construction purposes are increased. The mechanics of materials, which deals with structure of material and their reflections on external performance and properties of material systems, is rapidly developing. It involves the study of structural characteristics at several levels - from bond forces to the geometric arrangement of particles in a composite. Depending on the physical property being investigated, the desired characteristics have a microscopic or macroscopic expression.

4. Structural and building engineering

Concept design is the first design stage. They are preliminary studies whose purpose is to establish whether the project is viable, to assist in the development of the project brief and to aid the identification of feasible options. The preferred option is then being developed into a concept design which is a response to the project brief. The project brief will continue to develop as the concept design is prepared, but is then frozen at the end of the concept design stage and change control procedures are introduced. Concept design is followed by detailed structural design during which the design develops to describe all the main components of the building and how they fit together [5].

Structural and building design can be defined as a creative activity based on ability to meet scientific principles to technical and environmental requirements based on behaviour predicting of the structure based on knowledge of statics, dynamics, flexibility and strength, materials science and structures, evaluation methods and the like. It is important to perform physical-technical analysis of buildings and structures, diagnose of structure and building defects and solve the renovation of the buildings. The aim is to design and create a reliable economic and environmental friendly structure that will serve the required purposes.

Until the middle of the 19th century, the design of the structure was largely based on the art of intuitively determining the dimensions and placement of structural elements and details, as well as shaping the overall appearance of the structure of the building. Gradually better and more accurate understanding and knowledge of the principles influencing the behaviour of the structure, its elements and individual materials, design procedures became more and more scientific, based on a number of laws and regularities that were explicitly described by mathematical - physical computational models.

Regardless of the method used, regardless of whether "manual analysis" or any computer system has been applied, design is always an optimization process, which is usually accompanied by certain constraints, such as physical, environmental, social and psychological. It is up to the creativity of engineers and architects to deal with them. To find a rational and optimal way out where it does not seem to exist, especially in cases where the layman cannot see it.

Creative activity requires the conception of new, increasingly complex physical models, which to a greater extent and depth capture not only the structure but also the behaviour of the actual building structure. Simultaneously with the new concepts, the mathematical description is being improved, which brings the possibility of generalizing the achieved results.

Very important, in the conceptual design phase, is the creative interaction - cooperation between the architects and the civil engineers. Architects, engineers, computer and information systems can now create highly efficient flexible ones that can very quickly change configurations of variant solutions, modify the created details of the structure, compare the aesthetic and economic benefits of individual variants and gradually find the optimal final shape of the structure, resp. building.

The design process usually includes the following phases: initial thoughts and ideas, introduction of limiting factors, architectural drawings and engineering sketches, visualization, in some cases also animation, engineering analysis and reliability assessment of the designed object, its optimization, processing of drawing documentation, etc.

In the traditional design model the individual phases of creation follow in series and represent an inelastic, time - consuming rigid system. The information flow between the architect and the engineer is restricted and is limited to the gradual completion of the individual stage. Following the introduction of systems based on BIM (Building Information Modeling) and thanks to modern information technologies traditional procedures significantly modify. The transmission and exchange of information in individual phases is of the "on-line" type, the creation takes place in parallel - side by side, which significantly shortens the design time intervals. Compatibility and identity of electronically exchanged data and documents ensure high design quality. Traditional procedures devoted only 15 to 20% of the time to conceptual activities and the remaining time was needed to process the technical documentation. Correct implementation of BIM engineering allows up to approximately 70% of the total time to be used for conceptual creative activities, and the remaining 30% is spent on project development.

5. Conclusions

Many new scientific and technical problems arise because technical works have to operate in increasingly demanding conditions, with ever-increasing sizes and loads. The growing quantity of relevant quantities is growing into a new quality, which requires new theoretical and experimental approaches, more complex models, new methods and algorithms in design and assessment. Creative activity requires the conception of new, increasingly complex physical models, which to a greater extent and depth capture not only the structure but also the behavior of the actual building structure. Simultaneously with the new concepts, the mathematical description is being improved, which brings the possibility of generalizing the achieved results.

We can no longer be satisfied with the fact that any engineering construction is "only" reliable and efficient, but these properties must correlate with environmental requirements. We cannot look for environmental aspects only in nature, but the environmental principle must be reflected in all areas of building and structure design. In the current conditions of broad-spectrum requirements for knowledge and field of application, the civil engineer is often placed in the task of effectively participating in the creation of the material environment and its structures and building composition in all stages. The inability to give construction works, as well as their parts and details, an appropriate environmental quality, is often and inappropriately obscured by arguments that the current designer and builder must be primarily pragmatic, must meet the client's – investor requirements. He must think professionally, technically, economically, and thus he does not have the space and time to deal with the environmental impact of his work.

References

[1] J. Schlaich, "On the conceptual design of structures - an introduction," *Proceedings of the Int. Symposium Conceptual Design of Structures, University of Stuttgart Proceedings,* vol. I, Stuttgart, 1996, pp.15-25, 1996.
[2] P. Juhas, *Strategy of steel structures field development*, SvF TU Košice, pp. 11-18, 2001.
[3] I. Senitkova, M. Senitko, "Sustainability in buildings: Buildings for healthy living," *Proceedings of the Int. Symposium Beyond sustainability. Between best practice and utopia.* Eindhoven : University of Technology, pp. 432-438, 2000.
[4] I. Senitkova, "Sustainable building design," *Proceedings of the Int. Conference Clima 2000,* Napoli 2001. - Brussels: Rehva, pp. 79-86, 2001.
[5] J. Connor, and S. Faraji, *Fundamental of structural engineering*, Springer International Publishing Switzerland, (eBook), second edition, 2013, 2016.

A Generalized Approach to Estimating the Out-of-plane Buckling of Steel Sections with a Triangularly Corrugated Web

Sergey Kudryavtsev[1]

[1]Department of Building Structures and Soil Mechanics, Ural Federal University, 17a Mira Street, Yekaterinburg 620002, Russia

s.v.kudryavtsev@urfu.ru

Abstract. Building practice shows that I-sections consist of flanges and corrugated web could be quite effectively used for the construction of compressed structures. In the present work, an investigation into the behaviour and design of steel bending and compressed members with the triangularly corrugated web is carried out. Based on the result of the finite element analysis (FEA) conducted by the ABAQUS an attempt is made to generalize approaches in the evaluation of the stability of such structures. The necessity for this work is since, in the most building codes and studies, the issues of the global buckling of such structures are considered separately for beam structures and for structures at central and eccentrical compression. Most of the building codes do not provide any recommendations and requirements for the design of compression members with corrugated webs at all and only consider the case of bending. It is shown that beam and column structures with triangularly corrugated web works almost identically in case of out-of-plane buckling. Finally, is proposed an expression allow to calculate the critical force in flanges of the cross-section in case of out of plane buckling caused by transverse bending, axial of eccentrical compression uniformly, that could be useful for practical structural design.

1. Introduction

In recent decades steel sections with corrugated webs are found to be more and more widely applicable in construction and for this reason, they attract more interest among researchers and designers. Modern building codes make an afford to give the recommendations for the design of beam structures with corrugated web [1-4]. Currently, in the world practice, there are three main systems for corrugated webs: trapezoidal corrugations [5-8] (Middle East, USA, Japan, China, Europe), wavy corrugations [8-11] (Europe, Turkey, China), triangular corrugations [12-17] (Russia, Kazakhstan, Malaysia). The question of assessing the stability of thin-walled structures, which include any steel structures, and even more so thin-walled corrugated profiles, is often decisive in determining the possibility of their use and to determine their bearing capacity [18]. At the same time, for elements with corrugated webs, the flanges of which are also relatively thin-walled (with thickness from 6mm to 20mm, as a rule), several scenarios of buckling are distinguished, which should be taken into account and studied at the design stage. For steel I-section, the following are possible: overall buckling mode, that is topical for any steel structure of beam or column type, in which either the entire section is in a compressed state (the case of axial compression), or just a part of the section is in tension, and another is compressed (the case of bending and eccentric compression).

The issues of local stability of compressed flange and corrugated web are left outside from the scope. This paper presents an attempt to formulate a generalizing approach to assessing the global buckling of steel sections with a triangularly corrugated web, as the most promising both from the point of industrial production and from the point of their application in the broadest conditions of practical construction.

The necessity for this work is since, in the most building codes and studies, the issues of the global buckling of such structures are considered separately for beam structures and for structures at central and eccentrical compression. Most of the building codes do not provide any recommendations and requirements for the design of compression members with corrugated webs at all and only consider the case of bending [15]. However, building practice shows that such sections can be quite effectively used for the construction of compressed structures – see figure 1 and 2.

Figure 1. Portal frame made of members with the triangularly corrugated web.

Figure 2. The multi-storey framework of a residential building consists of compressed and bending structures with triangularly corrugated webs.

The present research deals with the global buckling effect of bending and axially compressed steel members with triangularly corrugated webs – see figure 3. The finite element (FEA) linear buckling analysis (LBA) by the software ABAQUS (2017) was carried out for various cases including different lengths of the structures and density of corrugation expressed as a ratio of height to length of half-wave of corrugation. During analysis were obtained the elastic critical moments and results were compared with estimations of the currently available design guidance [1-4, 12].

2. Design rules for steel sections with triangular corrugated web

Most of the modern design codes give recommendation only for steel members with trapezoidal or sinusoidal corrugated web [1-4]. Design features of triangularly corrugated web members are given in manual [12]. Since the corrugated web is not supposed to resist axial forces, what is frequently called "accordion effect", fully or partially compressed members with corrugated web are designed considering that only the flanges resist the longitudinal axial force [5, 12-17]. Thus, for the structure stability, the following condition should be satisfied

$$N_f \leq \varphi_f R_y \gamma_c b_f t_f \tag{1}$$

in which N_f is a force in a most compressed flange of the member, b_f and t_f are flange width and thickness respectively, φ_f is reduction factor for out-of-plane buckling, R_y is the material yielding strength, $\gamma_c = 1.0$ is working condition factor specified by [1]. It should be noticed that compressed force in a flange could be expressed as follows: $N_f = M/h_f$ for the bending member, $N_f = N/2$ for the axially compressed member, $N_f = M/h_f + N/2$ for the eccentrically compressed member.

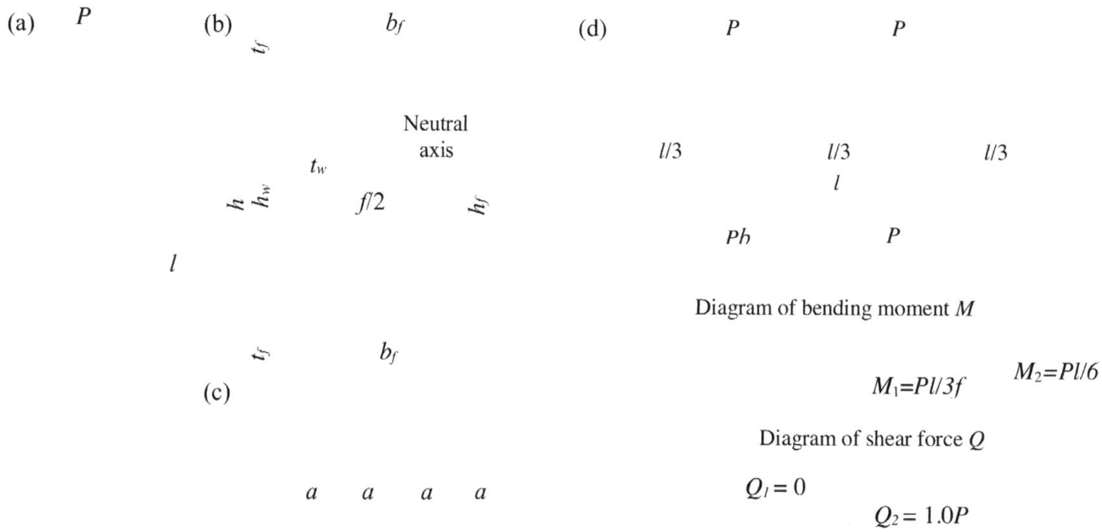

Figure 3. (a) Considered design scheme for axial compressed member; (b) Triangularly corrugated cross-section; (c) Corrugated web parameters; (d) Considered design scheme for bending member.

According to design code [1, 2], the value of buckling reduction factor in case of a member with corrugated web could be adopted as for the section with a flat web that can be described by following expressions

$$\varphi_f = 1 \text{ for } \bar{\lambda} < 0.6, \tag{2}$$

$$\varphi_f = \pi^2/\gamma_m \cdot \bar{\lambda}^{-2} \text{ for } \bar{\lambda} > 4.4, \tag{3}$$

$$\varphi_f = 0.5\left(\delta - \sqrt{\delta^2 - 39.48\bar{\lambda}^{-2}}\right)\Big/\bar{\lambda}^{-2} \text{ for other cases}, \tag{4}$$

in which $\delta = 9.87\left(0.96 + 0.09\bar{\lambda}\right) + \bar{\lambda}^{-2}$, $\bar{\lambda} = l/i_{min}\sqrt{R_y/E}$, l is a distance between anchorage points of the most compressed flange, i_{min} is a minimum value between the radius of gyration about strong $i_x = 0.5\left(h_w + t_f\right)$ and weak $i_y = 0.289b_f$ axis of corrugated web section, $\gamma_m = 1.3$ is a safety factor for elastic buckling.

3. Finite element analysis

While a numerical parametric study was considered two cases: axial compression and two-points bending of a member with corrugated web – see figure 3. All models are composed of two flanges 10mm x 200mm in size and 4mm x 600mm corrugated web. For the webs were assumed two different corrugation densities *f*/*a* = 50mm/100mm and *f*/*a* = 70mm/200mm. Such corrugation parameters represent practical geometries, which are commonly used for such structures in building practice [12-16]. A numerical study was considered four different lengths of compressed members *l* equal to 4m, 6m, 8m and 10m and seven different spans of bending members *l* equal to 3m, 4.5m, 5m, 6m, 7.5m, 9m and 12m. All compressed models were loaded by point axial force *P* = 1kN located at one end of the structure at the cross-section centre line. All bending models were loaded by two-point loads *P* = 1kN located in thirds of the span on the upper flange at the level of the structure centre line.

Boundary conditions were applied to either end of the models by restraining appropriate degrees of freedom to simulate the simply supported condition. An elastic material model was assumed with a yield strength value of 240MPa, modulus of elasticity E = 206000 MPa and Poisson's ratio 0.3.

Figure 4. The model adopted for the analysis of axially compressed member and its global elastic buckling mode.

Figure 5. The model adopted for the analysis of the bending member and its global elastic buckling mode.

4. Results of the numerical analysis

All models accepted for present work were buckled under perfect conditions, wherein no initial imperfections and eccentric load are considered. The general view of the global buckling mode of compressed and bending members are shown in figure 4 and 5 respectively.

The critical buckling load was calculated using the load factor obtained from the eigenvalue's analysis by multiplying it on the applied load. For example, *P* accepted for this study is 1kN, thus the critical buckling load $P_{cr} = 1\text{kN} \cdot v$, where v is eigenvalue received from the buckling analysis. In the considering case, a critical moment at global buckling of the bending element could be obtained as $M_{cr} = P_{cr} l/3$.

In such an approach, the reduction factors were calculated as a ratio M_{cr}/M_y for bending members and N_{cr}/N_y for compressed members. Results of FEA are presented in table 1 and 2.

Analyzing the data, it could be noted that for the relatively slender members with slenderness parameter $\bar{\lambda} > 4.4$ the minimum obtained values of reduction factors corresponds to the elastic buckling predicting by Euler's theory. For the relatively rigid members with $\bar{\lambda} < 4.4$, the minimum obtained values of reduction factors correspond to the inelastic buckling mode. In other words, members with corrugated webs behave like familiar structures with a flat web, but the first one has a greater load capacity as far as obtained reduction factors demonstrate the greater values for most cases up to 30% compared with the flat web structures of the same slenderness. As can be seen from the presented data the corrugation density weakly affects the buckling performance of the structure and values of the reduction factors for two types of corrugation looks almost similar. Moreover, it is distinctly that beam and column structures with triangularly corrugated web works almost identically in case of out-of-plane buckling. This fact once again confirms the "accordion effect" at which the corrugated web gets out from the work of the structure at bending (no matter longitudinal or transverse).

Generalizing the results, in figure 6 is given a graph shows the dependence of the values of reduction factors on the non-dimensional slenderness of member, that can be recommended for practical evaluation of load carrying capacity for steel triangularly corrugated web members.

Table 1. Results of linear buckling analysis (LBA) for bending members.

Model	$\bar{\lambda}$	M_{cr}, kNm	M_y, kNm	Reduction factor	
				M_{cr}/M_y	φ_f by Eq. (5)-(7)
f/a = 50/100, l = 3m	1.77	602.1	292.8	2.056	0.859
f/a = 50/100, l = 4.5m	2.65	298.3		1.019	0.713
f/a = 50/100, l = 5m	2.95	245.9		0.840	0.653
f/a = 50/100, l = 6m	3.54	177.1		0.605	0.535
f/a = 50/100, l = 7.5m	4.43	120.3		0.411	0.387
f/a = 50/100, l = 9m	5.31	89.2		0.305	0.270
f/a = 50/100, l = 12m	7.09	58.0		0.198	0.151
f/a = 70/200, l = 3m	1.77	631.8		2.158	0.859
f/a = 70/200, l = 4.5m	2.65	299.1		1.022	0.713
f/a = 70/200, l = 5m	2.95	246.2		0.841	0.653
f/a = 70/200, l = 6m	3.54	178.0		0.608	0.535
f/a = 70/200, l = 7.5m	4.43	121.1		0.414	0.387
f/a = 70/200, l = 9m	5.31	89.9		0.307	0.270
f/a = 70/200, l = 12m	7.09	58.5		0.200	0.151

Table 2. Results of linear buckling analysis (LBA) for axial compressed members

Model	$\bar{\lambda}$	N_{cr}, kN	N_y, kN	Reduction factor	
				N_{cr}/N_y	φ_f by Eq. (5)-(7)
f/a = 50/100, l = 4m	2.36	1670.2	960	1.740	0.767
f/a = 50/100, l = 6m	3.54	745.38		0.776	0.535
f/a = 50/100, l = 8m	4.72	420.67		0.438	0.341
f/a = 50/100, l = 10m	5.91	269.23		0.280	0.218
f/a = 70/200, l = 4m	2.36	1663.9		1.733	0.767
f/a = 70/200, l = 6m	3.54	744.3		0.775	0.535
f/a = 70/200, l = 8m	4.72	419.06		0.437	0.341
f/a = 70/200, l = 10m	5.91	268.64		0.280	0.218

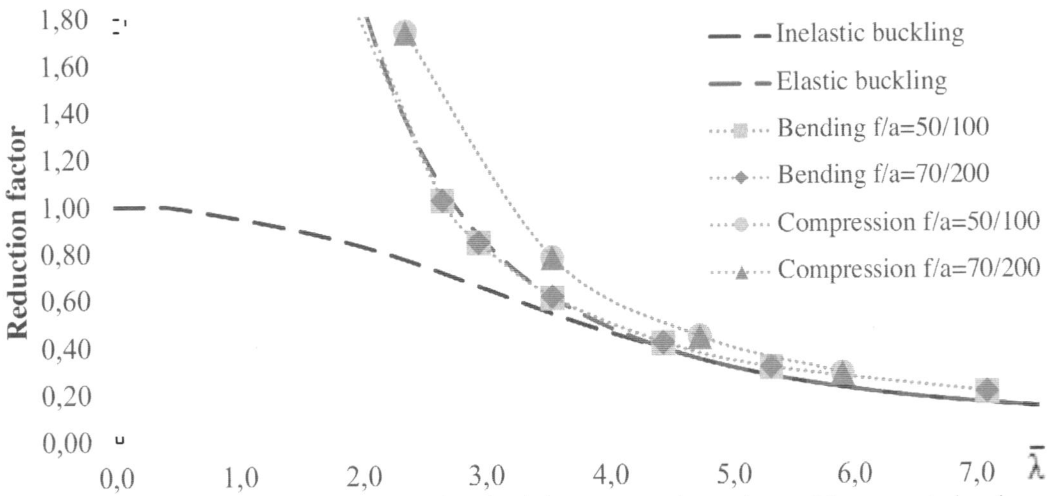

Figure 6. Buckling curves for bended and axial compressed members with corrugated web.

5. Conclusions

In the present work, an investigation into the behaviour and design of steel triangularly corrugated web bending and compressed members is carried out. Based on the result of the FEA an attempt is made to generalize approaches in the evaluation of the stability of such structures. Proposed expression (1) allows to calculate the critical force in flanges of the cross-section in case of out of plane buckling caused by transverse bending, axial of eccentrical compression uniformly, that could be useful for practical structural design.

Surely the buckling performance of such structures requires further investigation considering the influence of various boundary conditions, imperfections in corrugated web and its alignment relative the flanges.

References

[1] SP 16.13330.2017, *Steel structures*, Ministry of Construction of Russia, 2017, (in Russian).

[2] SP 294.1325800.2017, *Steel* Structures. Design Rules, Ministry of Construction of Russia, 2017, (in Russian).

[3] EN 1993-1-1, *Eurocode 3: Design of steel structures – Part 1-1: General rules and rules for buildings*, European Committee for Standardization, 2005.

[4] EN 1993-1-5, *Eurocode 3: Design of steel structures – Part 1-5: Plated structural elements*, European Committee for Standardization, 2006.

[5] E. Y. Sayed-Ahmed, "Design aspects of steel I-girders with corrugated steel webs," *El. J. of Structural Engineering*, vol. 7, pp. 27–40, 2007.

[6] Q. Inaam, and A. Upadhyay, "Flexural behaviour of steel I-girder having corrugated webs and slender flanges," *Structures*, vol. 27, pp. 12–21, 2020.

[7] B. Jager, L. Dunai, and B. Kovesdi, "Flange buckling behaviour of girders with corrugated web Part I: Experimental study," *Thin-Walled Structures*, vol. 118, pp. 181–195, 2017.

[8] H. Kazemi, "Lateral bracing of I-girder with corrugated webs under uniform bending," *J. of Constr., Steel Research*, vol. 66, Issue 12, pp. 1502–1509, 2010.

[9] G. C. Lopes, C. Couto, P. V. Real, and N. Lopes, "Elastic critical moment of beams with sinusoidally corrugated webs," *J. of Constructional Steel Research*, vol. 129, pp. 185–194, 2017.

[10] R. Aydin, E. Yuksel, N. Yardimci, M. H. Kisa, and T. Gokce, "In-plane behaviour of beam-to-column connections of corrugated web I-sections," *J. of Constr., Steel Research*, vol. 100,

pp. 183–196, 2014.
[11] Y. L. Guo, H. Chen, Y. L. Pi, and M. A. Bradford, "In-plane strength of steel arches with a sinusoidal corrugated web under a full-span uniform vertical load: Experimental and numerical investigations," *Eng. Structures*, vol. 110, pp. 105–115, 2016.
[12] S. Kudryavtsev, "*Calculation and design of welded I-section with corrugated webs,*" Yekaterinburg: Publ. of Ural University, 60 p., ISBN 978-5-7996-2056-1, 2017 (in Russian).
[13] S. Kudryavtsev, "Influence of web openings on bearing capacity of triangularly corrugated web beam," *IOP Conference Series: Materials Science and Engineering, vol. 365(4),* 042034, 2018.
[14] S. Kudryavtsev, "Lateral-Torsional Buckling Behaviour of Triangularly Corrugated Web Beamn" *IOP Conference Series: Materials Science and Engineering,* vol. 471, 052016, 2019.
[15] S. Kudryavtsev, "Buckling behavior of steel column with triangularly corrugated web," *MATEC Web Conf.*, vol. 279, 02007, 2019.
[16] S. Kudryavtsev, "Transverse load carrying capacity of steel triangularly corrugated web beam with opening," *MATEC Web Conf.*, vol. 279, 02004, 2019.
[17] F. De'nan, and N. S. Hashim, "Study on Bending Behaviour of Triangular Web Profile Steel Section by Finite Element Analysis," *Apl. Mech. And Mat.*, vol. 94-96, pp. 1539–1544, 2011.
[18] K. Murawski, 2020 Lateral buckling in elastic-plastic states of thin-walled semi-slender columns made of steel R35 according to known hypotheses *Preprint* 10.13140/RG.2.2.23578.59845.

IOP Conf. Series: Materials Science and Engineering 1066 (2021) 012003 doi:10.1088/1757-899X/1066/1/012003

Strength Verification of Structural Steels and Materials

Pavol Juhás [1]

[1] Institute of Technology and Business in České Budějovice, Department of Civil Engineering, Okružní 517/10, 370 01 České Budějovice, Czech Republic

15364@mail.vstecb.cz

Abstract. There is already a relatively wide range of structural steels and materials of various shapes and strengths for manufacturing of steel structures. They produce in different countries of Europe and the world. It is primarily a matter of the manufacturer, resp. the contractor of the steel structure with the vat and under what conditions he will provide the necessary construction materials. In addition to the price point of view, the quality of the considered construction materials is probably important. In particular, their geometric and strength suitability and accuracy are taken into account here. The quality of structural steels and materials is generally specified in the relevant standards and is the responsibility of the respective manufacturers. They declare it in the so-called quality certificates. At the same time, for various reasons, there may be a need to verify the declared quality, in particular the strength of specific structural steels and materials. The paper presents common probabilistic-statistical procedure and the strength procedures of structural steels and materials based on the test results, which are enabled by the current European standard EN 1990: 2002 as well as the transformed national standard CSN EN 1990: 2004, and similar other national standards. The practical application of the individual procedures and some details are illustrated by a specific example.

1. Introduction

Current standards for the design of steel structures contain a number of recommended structural steels of various strengths. These standards also state the basic strength characteristics of individual structural steels. These are mainly the yield strength f_y, resp. R_e, tensile strength f_u, resp. R_m and ductility A_5. The yield strengths and tensile strengths of the recommended structural steels depend on a number of random variables. For practical purposes, however, they are determined and standardized by specific values. The determination of the standard values of the yield strength f_y and the tensile strength f_u of the recommended structural steels and materials must therefore have a probabilistic basis [1-4].

Consistent probabilistic-statistical determination of the strength characteristics of structural steels requires large sets of the necessary data obtained by standardized tests. It should be noted that the probabilistic-statistically determined strength values of structural steels and materials are always only hypothetical, as they are determined on the basis of previous tests of already produced steels and materials for the design and construction of future structures. The determination of the strength of structural steels and materials by means of tests is enabled by the current European standard EN 1990: 2002 and transformed national standards, in CR it is CSN EN 1990: 2004. Effective determination of the strength of structural steels and materials is also enabled by simulation methods in which many new statistical sets of randomly variable strength quantities are created according to the chosen mathematically formulated type of probability distribution and their available information and statistical data obtained by tests. The Monte Carlo method is considered to be a very suitable simulation method for such purposes [5]. This paper provides basic information on determining the strength characteristics of structural steels and materials according to the indicated procedures and the relevance of statistical data. At the same time, these procedures are applied and compared in the evaluation of strength properties of selected structural steels S355 ML and S420 ML, which were used for the bridge Apollo in Bratislava [6].

2. Procedures for determining the strength of structural steels

Probabilistic-statistical procedure

Strength of structural steel expressed in general functional form

$$f(x) = f(x_1, x_2, x_3 \dots\dots x_n) = ([x]_1, [x]_2, [x]_3 \dots\dots\dots [x]_k),\ n = \sum_{i=1}^{k} n_i, \tag{1}$$

where x_1, x_2, x_3 to x_n, resp. $[x]_1$, $[x]_2$, $[x]_3$ to $[x]_k$ are individual random variables whose frequencies are n_1, n_2, n_3 to n_k, n is the number of all random variables of steel strength f_x determined by tests.

In general, for structural steels, tensile tests determine the yield strength f_y and the tensile strength f_u. For normative purposes and for the purposes of practical design, their characteristic values (determined with a probability of failure $P_f = 0.05$) and design values (determined with a probability of failure $P_d = 0.001$) are important. The characteristic values of the yield strength f_y and the tensile strength f_u of the structural steel are determined by the relations:

$$f_y = m_{fy} - \beta_{f,a} s_{fy} = m_{fy}(1 - \beta_{f,a} v_{fy}) \tag{2a}$$

$$f_u = m_{fu} - \beta_{f,a} s_{fu} = m_{fu}(1 - \beta_{f,a} v_{fu}) \tag{2b}$$

The design strength values of structural steel shall be determined using the relationships:

$$f_{yd} = m_{fy} - \beta_{fd,a}\, s_{fy} = m_{fy}(1 - \beta_{fd,a} v_{fy}), \tag{3a}$$

$$f_{ud} = m_{fu} - \beta_{fd,a} s_{fu} = m_{fu}(1 - \beta_{fd,a} v_{fu}) \tag{3b}$$

where m_f is the mean value, s_f is the standard deviation and v_f is the coefficient of variation of the relevant set of strength values f_x, a_f is asymmetry coefficient or slant [1,2].

Standard procedures

Depending on the relevance of the test results, the actual standards EN 1990: 2002, CSN EN 1990: 2004 and other assume the following two cases:

a) Only the results of a small number n of sample tests are available.
b) In addition to the results of the selection tests, previous knowledge providing a priori results is also available.

The first case is general and unambiguous - a, if only results from a limited number of n sample tests are available and no previous knowledge of the necessary statistical characteristics (mean value m_f, standard deviation s_f and coefficient of variation v_f) is available. In the standards it is recommended to use the second case - b, with a conservative estimate of the coefficient of variation v_f. In both cases, due to the number of tests, the assumption of a normal distribution or a two-parameter logarithmic-normal distribution of the occurrence of the evaluated strength is assumed. It is therefore not considered with a corresponding slant a_f, which, however, can be very significant, especially in a small number of tests [1, 2].

If normal distribution of the evaluated strength is assumed the characteristic values of the yield strength f_y and the tensile strength f_u of evaluated structural steel are determined:

$$f_y = m_y(1 - k_y v_y), \quad f_u = m_u(1 - k_y v_u) \tag{4a, 4b}$$

The design strength values of the evaluated structural steel are determined:

$$f_{yd} = m_y(1 - k_d v_y), \quad f_{ud} = m_u(1 - k_d v_u) \tag{5a, 5b}$$

The values of the coefficients k_y and k_d are contained in the standards EN 1990: 2002, CSN EN 1990: 2004 and other.

In accordance with the design method according to the limit states, the ratios of the determined characteristic values of the yield strength f_y or the tensile strength f_u and the design values of the strength f_{yd}, f_{ud} actually express the partial reliability coefficients of the evaluated structural steel γ_{M0} and γ_{M2}.

$$f_{yd} = f_y / \gamma_{M0}, \quad f_{ud} = f_u / \gamma_{M2} \tag{6a, 6b}$$

If the log-normal distribution of the evaluated strength is assumed the characteristic values of the yield strength f_y and the tensile strength f_u of the evaluated structural steel are determined:

$$f_y = \exp(m_{y,\ln} - k_y s_{y,\ln}) \cong m_y \exp(-k_y v_y) \tag{7a}$$

$$f_u = \exp(m_{u,\ln} - k_y s_{u,\ln}) \cong m_u \exp(-k_y v_u) \tag{7b}$$

The design strength values of the evaluated structural steel are determined:

$$f_{yd} = \exp(m_{y,\ln} - k_d s_{y,\ln}) \cong m_y \exp(-k_d v_y) \tag{8a}$$

$$f_{ud} = \exp(m_{u,\ln} - k_d s_{u,\ln}) \cong m_u \exp(-k_d v_u) \tag{8b}$$

The values of the coefficients k_y and k_d are contained in the standards EN 1990: 2002, CSN EN 1990: 2004 and other.

3. Strength evaluation of selected structural steels

The yield strength f_y and the tensile strength f_u of sheets with thickness $t \leq 40$ mm made of structural steels S355 ML and S420 ML are evaluated on the basis of a database of certificate results [5, 6]. The relevant standard characteristic and design strengths of the evaluated structural steels are:

Steel S355 ML $\rightarrow f_y = 355.00$ MPa, $f_{yd,EN} = 355.00$ MPa, $f_{yd,CSN} = 295.83$ MPa,
$f_u = 470.00$ MPa, $f_{ud,EN} = 427.27$ MPa, $f_{ud,CSN} = 361.54$ MPa.
Steel S420 ML $\rightarrow f_y = 420.00$ MPa, $f_{yd,EN} = 420.00$ MPa, $f_{yd,CSN} = 336.00$ MPa,
$f_u = 520.00$ MPa, $f_{ud,EN} = 472.73$ MPa, $f_{ud,CSN} = 400.00$ MPa.

The considered sets of $f_{y,i}$ and $f_{u,i}$ are contained in Table 1. The corresponding histograms are shown in Figure 1. The results of the statistical evaluation of the considered structural steels are in Table 2.

Table 1. Sets of the yield strength f_y and tensile strength f_u.

Steel S355 ML				**Steel S420 ML**			
f_y		f_u		f_y		f_u	
Classes	Count	Classes	Count	Classes	Count	Classes	Count
360	0	460	0	360	0	460	0
370	0	470	0	370	0	470	0
380	1	480	5	380	0	480	0
390	6	490	7	390	0	490	0
400	7	500	37	400	0	500	2
410	11	510	24	410	10	510	0
420	23	520	34	420	10	520	15
430	24	530	28	430	18	530	14
440	22	540	16	440	19	540	21
450	24	550	13	450	20	550	18
460	18	560	8	460	15	560	27
470	16	570	1	470	15	570	22
480	12	580	2	480	7	580	17
490	6	590	1	490	3	590	3
500	3	600	0	500	12	600	2
510	1	610	0	510	8	610	0
520	0	620	0	520	2	620	2
530	0	630	0	530	1	630	0
540	2	640	0	540	2	640	0
550	0	650	0	550	1	650	0
560	0	660	0	560	1	660	1
Total number of values n = 176				Total number of values n = 144			

Table 2. Results of statistical evaluation of the steel S355 ML and S420 ML.

Parameter	**S355 ML**		**S420 ML**	
	f_y	f_u	f_y	f_u
Number of values n	176	176	144	144
Average value m	439.13	516.31	454.35	549.49
Average value m_{ln}	6.0828	6.2459	6.1163	6.3081
Standard deviation s	27.997	20.879	32.741	23.566
Standard deviation $s_{ln,a}$	0.0632	0.0401	0.0708	0.0424
Standard deviation $s_{ln,b}$	0.0637	0.0404	0.0720	0.0429
Coefficient of variation v	0.0638	0.0404	0.0721	0.0429
Slant a	0.4334	0.5070	0.6707	0.6660
Minimum value f_{min}	379	474	401	498
Maximum value f_{max}	539	582	556	652
Difference of extreme values Δf	160	108	155	154

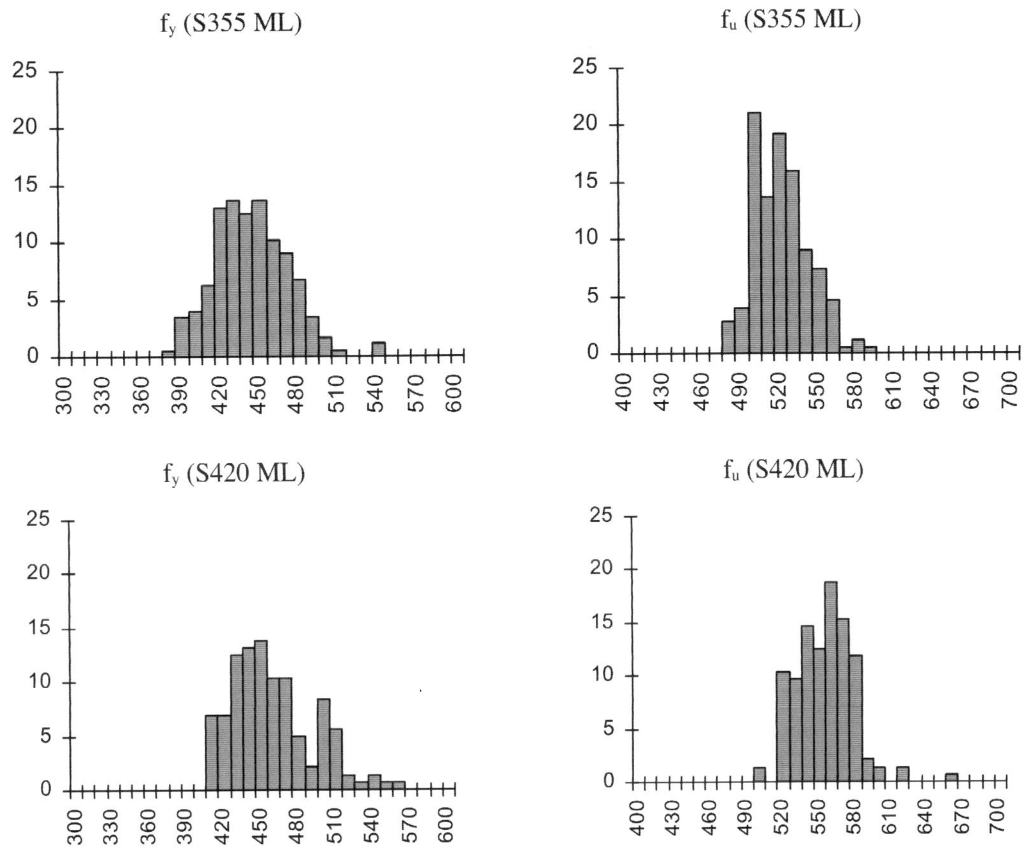

Figure 1. Histograms of the yield strength f_y and tensile strength f_u.

For both structural steels, characteristic and design values of strength f_y, f_u, f_{yd} and f_{ud} are determined according to the presented probabilistic statistical procedure (PP) and according to the standard procedure (NP). In the case of a probabilistic-statistical procedure, a normal distribution, a log-normal distribution and a gamma distribution of the evaluated strengths are assumed [1, 2]. The standard procedure assumes a normal and log-normal distribution of the evaluated strengths, a small number of n sampling tests is considered (case a). For comparison, the values of strengths f_y, f_u, f_{yd} and f_{ud} are also determined by the simulation method Monte Carlo (MC) using the calculation program MC-SIMUL [7]. The determined characteristic values f_y, f_u and design values f_{yd}, f_{ud}, as well as their mutual comparison are summarized in Table 3 and 4.

Table 3. Characteristic and design values of strength and their mutual comparison, S355 ML.

Procedure and expected distribution		f_y	f_u	f_{yd}	f_{ud}	f_y/f_{yd}	f_u/f_{ud}
		MPa					
PP	normal	392.90	482.10	352.56	451.86	1.114	1.067
	log.-normal	396.66	485.14	367.38	464.76	1.080	1.044
	gama	396.86	485.32	369.26	466.40	1.075	1.041
NP	normal	392.76	481.75	350.92	450.55	1.119	1.069
	log.-normal	394.70	482.75	359.12	454.67	1.099	1.062
MC	normal	399.78	486.76	355.08	458.25	1.126	1.062
	log.-normal	390.34	483.77	356.62	455.50	1.095	1.062

Table 4. Characteristic and design values of strength and their mutual comparison, S420 ML.

Procedure and expected distribution		f_y	f_u	f_{yd}	f_{ud}	f_y / f_{yd}	f_u / f_{ud}
		MPa					
PP	normal	400.46	510.71	353.12	476.65	1.134	1.071
	log.-normal	406.86	515.39	379.94	495.81	1.071	1.039
	gama	407.37	515.64	383.33	498.24	1.063	1.035
NP	normal	400.04	510.40	350.79	474.96	1.140	1.075
	log.-normal	403.02	511.69	362.33	480.05	1.112	1.066
MC	normal	417.74	520.39	373.89	478.71	1.117	1.087
	log.-normal	408.95	513.47	372.79	482.08	1.097	1.065

4. Discussion

In the presented evaluation, the current procedures for determining the characteristic and design strength values of the considered structural steels were applied. The obtained results do not show more significant differences between the individual procedures and the expected distributions of the random occurrence due to relatively homogeneous sets of the evaluated values of yield strength $f_{y,i}$ and tensile strength $f_{u,i}$. The obtained results also provide essential information about the basic strength properties of the assessed structural steels. They result in certain qualitative differences.

The determined characteristic and design strength values of structural steel S355 ML comply with the relevant standard values, resp. they even exceed them. In this context, when determining the design values of strength f_{yd} and f_{ud}, it is also possible to accept the values of partial reliability coefficients of the assessed steel $\gamma_{M0} = 1.0$ and $\gamma_{M2} = 1.1$ according to the actual standards for design of steel structures EN 1992-1-1: 2005 and CSN EN 1993-1-1: 2007. The coefficients $\gamma_{M0} = 1.2$ and $\gamma_{M2} = 1.3$ according to the previous standard CSN 73 1401: 1994 appear to be too safe for the evaluated structural steel.

The determined characteristic values of the yield strength f_y of the structural steel S420 ML are less than or close to the standard value $f_{y,N}$. The corresponding differences are insignificant. However, the determined design values f_{yd} are significantly smaller than the standard value $f_{yd,EN}$. Based on these results, it can be stated that the partial reliability factor $\gamma_{M0} = 1.0$ according to the actual standards EN 1992-1-1: 2005 and CSN EN 1993-1-1: 2007 is not safe. The achieved result is better satisfied by the partial coefficient $\gamma_{M0} = 1.10$ to 1.15.

The determined characteristic values of the tensile strength f_u of the structural steel S420 ML are close to the standard value $f_{u,N}$. The achieved result is satisfied by the partial reliability factor $\gamma_{M2} = 1.1$ according to the actual standards for the design of steel structures EN 1992-1-1: 2005 and CSN EN 1993-1-1: 2007. The coefficient $\gamma_{M2} = 1.3$ according to the previous standard CSN 73 1401: 1994 appear to be too safe for the evaluated structural steel.

5. Conclusion

Determination and verification of strength characteristics of structural steels and materials on the basis of relevant test results and presented procedures according to actual standards is justified and necessary especially for generally demanding structures.

The results of the evaluation of the strength properties of the assessed structural steels S355ML and S420ML show certain qualitative differences. In this context, the uniform values of partial reliability coefficients γ_{M0} and γ_{M2} ($\gamma_{M0} = 1.0$ and $\gamma_{M2} = 1.1$) promoted for all structural steels of different strengths appear to be unjustified.

IOP Conf. Series: Materials Science and Engineering 1066 (2021) 012003 doi:10.1088/1757-899X/1066/1/012003

The results of the presented evaluation encourage caution in the application of partial reliability coefficients of structural steels and materials according to actual standards for the design of steel structures.

References

[1] A. Mrázik, *Theory of reliability of steel structures*, VEDA, Bratislava 1987.
[2] P. Juhás, et al., "Design of steel structures," *Commentary on STN 73 1401:1998.* SÚTN, Bratislava 2001.
[3] V. Raizer, *Reliability of structures – analysis and application*, Backbone Publishing Co., New York 2009.
[4] P. Juhás, *Reliability and strength of materials of steel structures*, TU Košice, ELFA, Košice 2009.
[5] P. Marek, et al., *Simulation based reliability assessment for structural engineers*, CRC Press, Boca Ration/New York/London/Tokyo 1995.
[6] K. Kálna, and M. Vitásek, *Welding procedures setting and testing of welding connections properties of Kosicka (Apollo) bridge*, Welding 53, No 5, 2004.
[7] E. Kulkova,"Application of Monte Carlo method applied to load effect combinations," *Proceedings of Int. Conf.on Metal Structures,* Politechnika Krakowska, Crakow, pp.159-168, 1995.

Construction-Technical Survey of a Historical Building in Loket

Ales Kankovsky[1], Martin Dedic[1]

[1] Institute of Technology and Business in České Budějovice, Department of Civil Engineering, 370 01 České Budějovice, Czech Republic

kankovsky@mail.vstecb.cz

Abstract. The main objective of the article is to inform about the construction-technical survey, which focuses on the existing historical building from the first half of the 19th century, in Loket city, Czech Republic. In the past, the building was hit by floods, which affected its technical condition, the deterioration of which also affected the subsequent abandonment of the building, and neglect of maintenance. The main research issues stem from the building-technical survey focused on the condition of structures, especially the load-bearing wall structures, the ceiling structures, and the roof structures. After detecting the faults and other possible damage, we look for their causes and find out which of them are fatal to the building and which can easily remove. Of the evaluated structures, we focus mainly on their damage due to moisture, excessive structures loading, and changes in foundation conditions caused by previous floods. Furthermore, we assess the impact of failures on the structure and operation of the building. In conclusion, we propose measures based on the structure´s found failures and possibilities for their repair.

1. Introduction

The task was to determine the technical condition and extent of failures in structures affected by previous floods, i.e. primarily the perimeter and internal load-bearing masonry, especially in the building basement. The survey focused on the damage extent and overall technical condition of the truss structure degraded due to neglected maintenance. The main goal of the construction-technical survey is to locate faults and other possible damage, to determine their extent and determine their cause, as well as to assess the possibilities of eliminating these faults and damage.

The solved object is composed of three rectangular tracts (1x 6.7 * 11.85 m, 2x 4.75 * 9.90 m), while the larger central part (risalit) is three-story and the smaller outer parts are two-story. The building is also partially basement from the imaginary longitudinal half of the building in its western, southwestern, and northwestern parts. The built-up area of the building is 175 m2, the maximum height of the building from the adjacent terrain is up to 15 m, the depth of the basement is up to 2.0 m below the level of the adjacent terrain. The ground floor of the building is located above the level of the adjacent terrain. The construction system of the building is walled, made of bricks or stone. Ceiling structures are concrete or wooden formed by a semi-combustible beamed ceiling. The individual tracts are covered by a hip roof with roofing made of profiled sheet metal, under which is located the hydro isolation foil. The trusses are wooden, with massive purlins and joists. The historic façade is characterized by a massive plastered plinth, strip bossage to the height of the second floor, profiled shutters of windows and doors, window ledges, parapet cornices, strip (cordon) cornice, and main crown (drip) cornice. The initial use of the building was living in a family house (villa). In history, the building has changed its purpose several times and served as a school canteen and then as a police station. Subsequently, the building was abandoned for a long time and these days it is not used. In 2009, the town of Loket was hit by a flood, which flooded the basement of the building. [1]

2. Construction-technical survey

We have carried out two local surveys as part of the solved building construction and technical survey (local survey dated to 21st of May 2020 and local survey dated to 14th of July 2020), during which a survey of the building and the surrounding area was carried out, including photo documentation. On selected structures were also performed Non-destructive perceptual tests. Based on the found obvious faults, probes were performed in selected places of the building to verify the existing structures and determine their technical condition.

2.1. Cracks in the joints of perimeter masonry:

During the local survey, visual inspection revealed cracks in the internal and external plasters at both two-story outer parts of the building, on both floors, at the windowsill, lining, and lintel of the window, leading through the windowsills and window edging. Cracks are located in the basement and non-basement parts of the building. Inside the building, at the point of the window sill, cracks are visible at the point of contact between the sill and the window lining. Based on the survey a probe was performed (removal of the internal plaster of the lintel and parapet and discovery of the supporting structure) at one of the affected places. After removing the surface layer a crack was found in the place of the window sill in the vertical and horizontal joints of cross-walled brickwork, characterized by chipped mortar up to 3 cm deep. In the case of the window lintel, no crack was found in the joints of the masonry vault, or through the brickwork. After removing the plaster of the window lintel, the degraded condition of the original masonry mortar was also revealed, which crumbled at the edge of the bricks. The cracks on the façade and inside the building are shown in Figure 1 and Figure 2.

Figure 1. The western and eastern facades of the building with the marked places and direction of the cracks.

Figure 2. The probe in the windowsill and lintel on the 2nd floor of the building east wing.

2.2. Humidity in the basement:

In 2009, the town of Loket was affected by floods, which, due to the location of the solved building, caused the flooding of its basement in its entirety. The flooding of the basement caused an increase in humidity in the foundation structures, in the perimeter and interior walls, and in the ceiling structures. Due to the long-term effects of moisture, there was a gradual degradation of the internal plaster walls and ceilings, and partial disintegration of the brickwork, incl. connecting mortars, in exposed places without plaster (especially at corners and lintels of openings). Insufficient drying of wet structures subsequently caused the formation of mold and salt blooms. However, the perimeter and interior walls, including the vaulted ceiling structures, do not show signs of mechanical damage in the form of statically severe cracks, greater disintegration of the material, etc., which would disrupt the stability of the building or limit its use. The extent of basement damage due to long-term exposure to moisture is shown in Figure 3.

Figure 3. The extent of basement damage due to long-term exposure to moisture.

2.3. Rotten and damaged wooden truss elements:
During the local survey, a visual inspection revealed a neglected condition of truss parts, containing deposits of bird droppings, remnants of plaster and building material, places with rot on the truss foundation and adjacent wooden structures (tie beams, wall beam, rafters), including different coloring of elements. At the same time, rotten heads of a part of wooden tie beams were found, especially in the place of the perimeter wall. A sound test in the form of tapping the individual wooden elements of the truss revealed dull and hollow sounds emanating from some elements, which may indicate biological damage by wood-destroying insects. Visual inspection confirmed traces of deep scratch caused by wood-destroying insects (especially in the rafters of the truss in the central building part). [2] The extent of wooden truss elements damage is shown in Figure 4.

Figure 4. The extent of wooden truss elements damage.

2.4. Verification of ceiling structures
Verification of the real state of the existing wooden ceilings above the 1st and 2nd floors, which is shown in Figure 5, arises due to the cracks on the façade and due to the layout changes proposed by the investor, considering the future construction of new brick partitions.

Figure 5. The technical condition verification of the ceiling structure.

3. Results and discussions
3.1. Cracks in the joints of perimeter masonry:
Cracks in the joints of the perimeter masonry in the area around the window openings (lintel, lining, window sill) at both extreme parts of the building can be caused by two possible causes.

a) The building foundation
The first possibility is a change in the soil conditions in the basement of the building, caused by the mentioned floods in 2009. As these are old cracks in the vertical and horizontal joints of the masonry, these cracks indicate a gradual decline of the structure (foundation), see figure below.

b) Structure overload
The second possibility, due to cracks in the place of the window lintel, is the overload of the lintel by the wooden ceiling, or the ceiling beam placed in the place above the window lintel. Due to the fact that the building has changed the purpose of use many times in the past (housing, civic amenities), there were changes in the values of the live load, while increasing these live load values could lead to the overloading of the ceiling structure. Due to the subsequent distribution of forces, it decreased in the weakened place of the lintel, which caused cracks in the brick vault, which were reflected in the plaster at the junction of the lintel and the window lining. With a slight decrease in the window lintel, the ceiling structure in the place where the beam was placed probably also decreased, which also caused a decrease in the window sill above the ceiling structure due to the occurrence of deformations.

3.2. Humidity in the basement:
Due to the flooding of the basement in the past (floods in 2009), the current situation is most likely caused by insufficient drying of the premises after water depletion. According to the previous operators of the building, there was no waterlogging of the foundation/basement structures due to ground moisture. Whether the building has a waterproofing layer (original or additional) is not known.

3.3. Rotten and damaged wooden truss elements:
In the past, the roof most likely underwent a partial reconstruction, during which the roof covering (currently profiled sheet metal) was replaced and a waterproofing under-roof layer - foil - was additionally used in the truss of the building south wing. The attacked wooden truss elements are located in the parts of the north wing and the central part, without the hydro insulating foil. For this reason, the cause of the disintegration of the tops of wooden tie beams is mainly moisture in the structure due to leakage into it, which in long-term action, together with suitable temperature conditions and truss maintenance, causes wood-destroying rot, wood-destroying fungi and wood-destroying insects. As some wooden elements have deep surface scratches of a specific form, these may be ingestions caused by beetles "*Corymbia Rubra*" or "*Lyctus brunneus*".

3.4. Verification of ceiling structures
The technical condition verification of the ceiling structure of the wooden semi-combustion ceiling with the wooden flap and embankment was based on the values measured from the performed probe during the local survey. However, there were no defects or damage at the probe site. Also, frequent changes in the use of the building and thus changes in the live load could affect the reduction of the mechanical properties of wooden elements or could cause their local damage.

4. Conclusions

4.1. Cracks in the joints of perimeter masonry:
In the case of cracks in the structures, it is necessary to measure them and find out whether the cracks are active or passive. On the crack, we monitor its age, shape, and direction, the degree of opening, and the change of the crack over time. For this reason, we recommend the detection of all cracks on the internal and external plaster, in the presumed cracks, and subsequent observation of crack activity by applying a mortar/plaster target, which in case of failure indicates that the cracks are active and the process still continues.

Passive cracks can be sealed or injected using cement mortar, cement laitance and concrete mix, or modern technologies based on the principle of synthetic substances. Repair of Local and active wide cracks can be by stitching using steel clips, or sheathing (shotcrete). At the same time, it is possible to fix local cracks and additional reinforcement of structures by gluing steel reinforcement into the prepared grooves in the joints of the structures / drilled holes or their combination (HELIFIX system). This method can also be considered in the case of loose bricks, separated layers, etc. (e.g. in the case of lintels of windows and doors).

4.2. Humidity in the basement:
Residual moisture in the cellar, if it is not rising/seeping ground moisture, can be removed by sufficient drying of the premises and subsequently proposed remediation. Drying of unwanted moisture can be done in several ways, and construction interventions are very limited here, so remediation is limited primarily to the principles of physical and chemical.

In the case of Physico-chemical principles, it is possible to consider grouting masonry by a professional company, which will choose the most suitable solution (method of design and material) based on the actual condition. However, this type of measure is costly. The most available effective measures appear to be physical principles, such as intensive ventilation of cellars and the use of electric dehumidifiers (ideally on a condensing basis). If the impact of cellars with moisture is extensive to fatal, without long-term measures (e.g. after flooding), invasive measures in the form of installation of intensive heat sources together with effective dehumidifiers should be considered. This is a one-off measure, but it is also costly in terms of energy consumption. Other physical measures in the form of microwave wall heating and electroosmosis can also be considered.

4.3. Rotten and damaged wooden truss elements:
Excess moisture in the affected wooden truss elements can be additionally removed by thorough drying or complete replacement of the elements, depending on the extent of the damage. The rotten head of the tie beams must be replaced. Depending on the extent of the damage, wooden elements infested with fungi / wood-destroying insects can also be rehabilitated by impregnation methods of the infested wood (in the case of a greater extent of damage, efficiency decreases), as well as microwave remediation, hot air remediation or grouting of wood elements. However, these are costly methods. For a larger extent of damage to more elements of the recommendation, we recommend replacing all elements in the truss.

4.4. Verification of ceiling structures
Reinforcement of the existing wooden ceiling structure can be done with the help of additional support of the existing beams, in the place of the assumption of increased static load, with the enclosed steel profiles. The dimensions of individual additional elements must be chosen on the basis of detailed verification of the technical condition of existing structures and subsequently a detailed static calculation.

4.5. Recommendations for further survey activities
Due to the age of the building, its long-term non-use, and due to the occurrence of several types of defects and structural failures, we recommend performing:

- Removal of plasters in the entire extent of the basement space in order to determine the extent of damage due to moisture, or due to static load action (settlement of subfloors, loading, etc.).
- Carry out humidity measurements of the affected wall and ceiling structures in the cellars.
- In the place of cracks on the facade of the building, perform a complete removal of the plaster inside the building, in the affected places around the window openings, and finding out the actual extent of damage by static load cracks.
- Carry out crack measurements using a plaster target.
- Carry out a diagnosis of wooden elements in the truss, find out the actual extent of damage to elements exposed to moisture / wood-destroying insects, fungi, and rot.
- Revealing the ceiling structures in place of the designed partitions in order to determine the actual condition of the ceiling beams as part of the verification of their load-bearing capacity.
- Creating a digital model of a building using a 3D scanner or digital photogrammetry for accurate measurement of failures in one time period. It is important to choose the right LOD before starting work [3,4,5].

Acknowledgment(s)
This work was supported by the Grant Agency of the Institute of Technology and Business in České Budějovice, grant No. IGS_8210-015.

References
[1] M. Dědič, "Analysis of Historical Residence on Terms of the Current State," *IOP Conference Series: Material Science and Engineering*, vol. 471, 2019.
[2] M. Dědič, "Defects of roofing and roof structure – Apartment house Kaplice," *IOP Conference Series: Material Science and Engineering*, vol. 603, 2019.
[3] K. Prušková, M. Dědič, and J. Kaiser, "Possibilities of Using Modern Technologies and Creation of the Current Project Documentation Leading to the Optimal Management of the Building for Sustainable Development," *Central Europe towards Sustainable Building (CESB19)*, 2019.
[4] K. Prušková, "Building information management: Knowledge management principles used in the whole building lifecycle within the building information modeling technology", *SGEM 2018*, Albena, Bulgaria, vol. 18, Issue 2.1, pp 209-215, 2018.
[5] M. Dědič, "Evaluation of the processes of creating a project documentation of an existing building using a 3D scanner", *SGEM 2019*, Albena, Bulgaria, vol. 19, Issue 2.2, pp 127-132, 2019.

IOP Conf. Series: Materials Science and Engineering 1066 (2021) 012005 doi:10.1088/1757-899X/1066/1/012005

Ensuring Blast Resistance of Critically Important Buildings and Constructions in Case of Air Crash

A A Komarov [1], N V Gromov [1], O N Korolchenko [1]

[1] National Research Moscow State University of Civil Engineering, Yaroslavskoye Shosse, 26, Moscow, 129337, Russian Federation

ikbs@mgsu.ru

Abstract. Introduction. The paper is focused on improving blast resistance of critically important buildings and constructions exposed to a deflagration explosion due to an aircraft (A/C) crash in their vicinity, by improving methods of the explosion effects calculation. The main tasks include: 1. considering the physical processes that occur during an aircraft crash, which causes the formation of a highly explosive cloud; 2. quantifying the parameters affecting explosion loadings; 3. provide insights into modern procedures for calculating effects of a deflagration explosion caused by an A/C crash; 4. giving an example of calculating the loadings on a critically important building or construction in a crash deflagration explosion. Methodology and calculations. The article presents a methodology for calculating the maximum parameters of explosion loadings on buildings and constructions arising from aircraft crash situations. The calculations are performed with the help of author's and generally accepted methods by the numerical method with the use of MatLab software complex. Results and discussion. As a result of calculations, the proposed methodology allows to obtain: the fuel mass values capable of forming highly explosive mixture; to choose the appropriate scenario for the development of an aircraft crash: an igneous ball or a deflagration explosion of a fuel-air mixture; to determine the time dependence of the fuel vapor concentration in the air; to evaluate the maximum apparent flame front speed; to determine the dynamic parameters of the fireball and the time dependences of the overpressure at points in the space adjacent to the explosion site; to build up a maximum pressure field created by this deflagration explosion; to obtain the explosion loading integral parameters: the maximum and minimum explosion overpressure, the compression phase pulse, the probability of destruction of buildings; to evaluate vibration loading on the building from a deflagration explosion. Conclusion. The methodology presented in the article can be used to calculate the loadings on buildings and structures during a deflagration explosion that occurs when an aircraft crash.

1. Introduction

Aircraft crashes, despite their low probability, pose a great danger to any facilities located near the aircraft crash site. This is due to the presence on board of any aircraft of a large amount of fuel in a liquid or gaseous state. The damage effects in an aircraft crash, in addition to the impact of a glider or engines, are the fire of the jet fuel spillage, heat loadings from an igneous ball, as well as loadings from a compression wave during a deflagration explosion of fuel vapor.

The loadings arising from an aircraft crash are not of particular interest for civilian facilities due to the low probability of such an event. However, when designing such critically important constructions as nuclear power plants (NPPs), it is necessary to determine the parameters of explosion loadings that occur when an aircraft crashes within the NPP site [1]. At the same time, the most conservative ideas about the development of an aircraft crash are used [2]. In this regard, they consider the crash within the site of a NPP of a large commercial aircraft such as Boieng-747, Airbus A-320 or Antonov An-124 [3, 4].

Such regulatory documents as RD 03-409-01 "Methodology for assessing the consequences of crash explosions of fuel-air mixtures" or RB G-05-039-96 "Guidelines for analyzing the danger of crash explosions and determining the parameters of their mechanical impact" use the TNT equivalent as the basis for calculating loadings from deflagration explosions, which significantly overestimates the calculated loadings [5–7]. Some aspects of crash deflagration explosions were considered in works [8–12], but they did not fully reveal the processes of formation of a fuel-air explosive cloud in the atmosphere and the effect of its explosion on the construction design of buildings and structures.

This article describes the basic principles and some research results of this problem. A large commercial aircraft — Boieng-747 is considered as a source of danger. The paper is aimed at improving blast resistance of critically important buildings and constructions exposed to a deflagration explosion due to an aircraft (A/C) crash in their vicinity, by improving methods of the explosion effects calculation. The main tasks include considering the physical processes that occur during an aircraft crash, which causes the formation of a highly explosive cloud; quantifying the parameters affecting explosion loadings; provide insights into modern procedures for calculating effects of a deflagration explosion caused by an A/C crash; giving an example of calculating the loadings on a critically important building or construction in a crash deflagration explosion.

2. Methodology and calculations

Let us consider the physical processes that occur during a plane crash, which causes the formation of a highly explosive cloud.

The following conditions are necessary for an explosion: fuel vapor formation with a combustible liquid, the mixing of the formed vapor with air in certain proportions, and the appearance of an ignition source.

When a plane crashes, the main source of fuel vapor is the heated parts of its engine. We assume that all the aircraft engine parts can be heated to an average of 460 °C. The engine weight of Boieng-747 is 4,300 kg. Then the total heat of one engine will allow evaporating no more than 1,520 kg of fuel. The alleged crashing aircraft has 4 engines. Therefore, when the plane crashes, the maximum amount of fuel capable of forming a fuel-air explosive cloud will be no more than M_f = 6,080 kg [6].

Some specialists use the amount of vapor that can be in an explosive state in the aircraft fuel tanks as the basis for calculating the potential explosive hazard of an A/C. Given that the total fuel tank capacity of a heavy commercial aircraft is about 300 m^3, we obtain that no more than 100 kg of fuel can be involved in explosive combustion with this approach. However, in conservative calculations of the potential explosive hazard of an A/C crashing within a nuclear power plant, it is necessary to take into account the amount of vapor that can form when it evaporates from the red-hot parts of the engine. This amount of vapor is much greater than can be found in the empty fuel tanks of any A/C.

The ignition of the resulting cloud of explosive mixtures is possible from the hot parts of the same engine or from sparks that occur at the moment of the impact. For example, it is known that a crash or a hard touchdown of an aircraft on soft ground is not always accompanied by deflagration and fuel explosion.

During the deflagration of the cloud formed, it creates an overpressure caused by the rapid heat release and combustion products, a fireball (area of heated combustion products) is formed, which rises quickly enough under the influence of the Archimedean force. It should be noted here that there are two different concepts: "fireball", which is an upward-moving mass of explosive combustion products, and "igneous ball", which is a surface burning mass of re-enriched fuel mixture. In the first case, an explosion of an explosive mixture occurred, accompanied by overpressure, and in the second case, surface burning of the re-enriched mixture took place, which is not accompanied by a noticeable overpressure. At the same time, an outside observer can classify both phenomena as two identical ones. This is due to the fact that deflagration time of 100 ms or a surface burning time of several seconds is equally fast for a person, and for a pressure propagating at a speed of about 340 m/s, these are two fundamentally different processes. In the first case, blast pressure arises, since the explosion products "did not have enough time" to leave the fire location, and in the second case, there is practically no pressure, since the explosion products "managed" to leave the fire location.

Therefore, in a scenario involving the appearance of an ignition source immediately after the plane crash, an igneous ball emerges, which is formed as a result of burning fuel vapor and dispersed drops of kerosene, which occurs as the fuel mixes with air. This scenario for the development of the accident will conditionally be called the first one.

With a certain delay of the ignition source (for a few seconds), a fuel-air mixture forms that is capable of exploding, which is accompanied by overpressure of significant intensity. In other words, when a source of ignition manifests itself with a certain delay, the deflagration explosion of a part of the fuel in an explosive concentration first occurs, and after that (with the supply of oxygen) the remaining fuel burns out in the form of an igneous ball and a spillage fire. This scenario for the development of the accident will conditionally be called the second one.

Let us take a closer look at the second scenario of the crash. It was previously shown that the maximum amount of vapor that can be generated by the heated parts of one engine of the aircraft can be no more than M_f = 1,520 kg.

Let us calculate the formation process of a highly explosive cloud. We will consider the most unfavourable scenario for the development of the crash.

Figure 1 shows the lines of equal concentrations of the mixture 10 seconds after the vapor release. The isolines correspond to the following volumetric concentrations of the resulting vapor-air mixture: 1% vol. (outer isoline), 2%, 5%, and 10% vol. The situation was calculated when at the initial moment of time, there was a mixture of fuel vapor and air, that is, at the initial moment of time there is not a mixture of vapor and air fragments near the crashing engines, but an "enriched" mixture with a volume concentration of 27%.

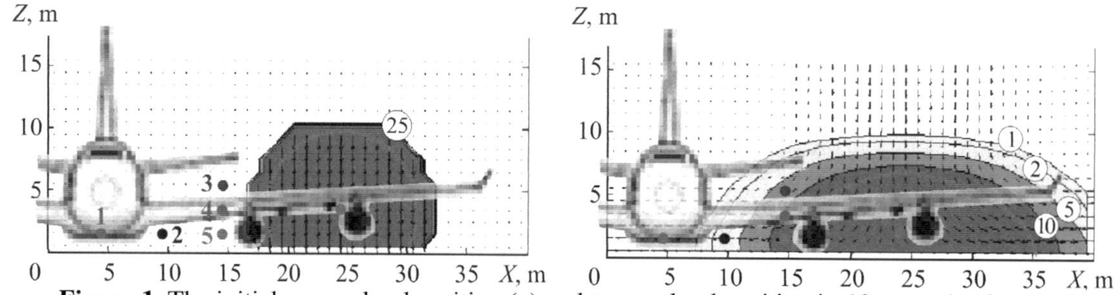

Figure 1. The initial vapor cloud position (a) and vapor cloud position in 10 seconds after vapor discharge (b).

When performing the calculations, it was assumed that there is an intensive mixing of fuel vapors with air, that is, the crash is accompanied by significant gas-dynamic flows and the crash occurs at the high values of the turbulent diffusion coefficient. The turbulent diffusion coefficient was assumed to be 0.5 m^2/s, which corresponds to the values observed in similar situations. In the figures, the green dots indicate the initial discharge limits, the position of which was determined for each moment. This allows us to judge the degree of deformation of the initial cloud due to gas-dynamic flows accompanying the aircraft crash. Besides, the figures show vectors of flow velocities.

Figure 2 shows the time dependencies of vapor concentration at five points, the position of which is shown in Figure 1.

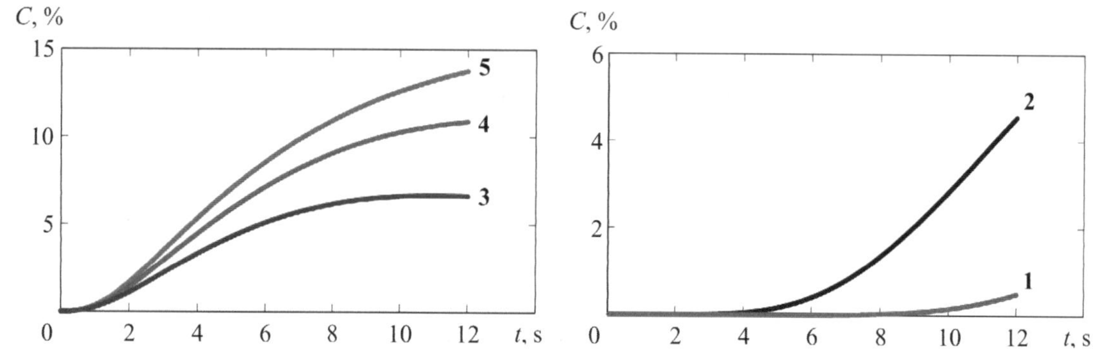

Figure 2. Vapor concentration time laws at five points.

Figure 3 shows the time law of the vapor mass having a concentration in a mixture with air from 1% vol. to 5% vol.

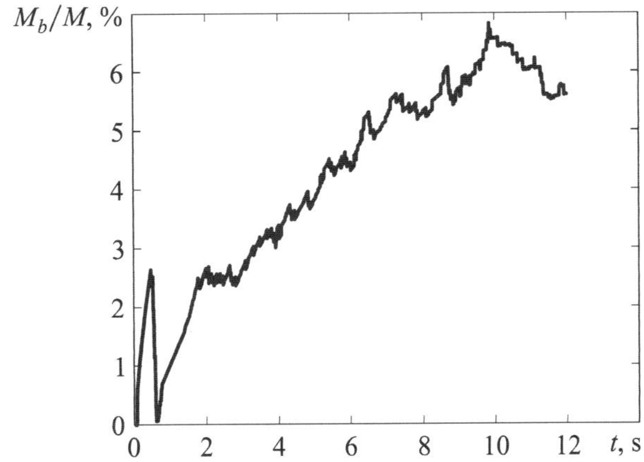

Figure 3. The time law of vapor mass ratio with volume concentration from 1% vol. to 5% vol. (M_b) to the total vapor mass (M).

The calculation results analysis showed that the initial conditions (the presence or absence of a large building near the site where the aircraft crashed; the initial state of the discharge: the discharge consists of vapor fragments or is already well mixed, etc.) have a negligible effect on the initial stage of the highly explosive cloud formation. In all the cases or in all the scenarios of the development of the aircraft crash causing a deflagration explosion at the time of the crash (the second scenario of the development of the aircraft crash), the maximum amount of vapor capable of creating conditions for deflagration does not exceed 7% of the initial vapor discharge mass. Given that the maximum amount of vapor that can be generated by the heated parts of one aircraft engine is not more than M_b = 1,520 kg, we obtain that the maximum vapor mass capable of involving a deflagration explosion, developing according to the second scenario, can be no more than M = 106.4 kg.

For the obtained value of the maximum vapor mass capable of exploding, we obtain the maximum radius of the fireball. Let us remember that that the fireball is the area of the explosion products and, accordingly, is determined by the boundary of the explosion products, and the igneous ball is the area of burning of the "enriched" mixture that burns as soon as the oxygen contained in the air is available. The weight of 106.4 kg of vapor (under normal conditions) is 16.45 m^3. The explosive mixture will have a volume of 822.5 m^3. Accordingly, the explosion product volume will be about 6,580 m^3. Assuming that the explosion products at the time of explosion completion will be in the form of a half cylinder mounted on the ground, we obtain that the maximum radius of the fireball will be no more than R_S = 19.5 m.

In order to determine the maximum apparent flame speed occurring during an explosion, it is necessary to determine the radius (length) at which the initial acceleration of the combustion process occurs due to the hydrodynamic instability of the mixture. Taking into account that a part of the combustion products (about 50%) will be required to expand the area of initial vapor discharge, we obtain that the main acceleration of the flame is carried out only at the last 4.9 m from the fireball boundary.

Based on the obtained values of parameters characterizing a fireball capable of forming in the process of a deflagration explosion, let us estimate the maximum apparent flame speed.

Following the recommendations of RD 03-409-01, we obtain that the maximum value of the apparent flame speed in deflagration explosion M = 106.4 kg of fuel will not exceed: $W = 43M^{1/6} =$ $= 93.6$ m/s.

Following the methods adopted in [13-19], the maximum visible flame speed in a deflagration explosion is defined as the product of the apparent flame speed performed under the conditions of hydrodynamic instability of the mixture (or turbulent combustion velocity) W_P, and the coefficients characterizing the way the mixture is initiated (k_1) and the nature of the surrounding area where the explosion occurred (k_2), that is $W = W_P\, k_1\, k_2$. Expert coefficient values k_1 and k_2 are given in [16,17].

Using an empirical ratio to determine W_P [16, 17], we obtain that

$$W_P = U_N\left(1 + A\sqrt{\frac{R_{PX}}{\Delta}}\right) = 11.3\ \text{m/s}, \tag{1}$$

where U_N — laminar apparent flame speed (U_N = 3.67 m/s); A — empirical proportion coefficient ($A = 0.34$); Δ — average size of detonation cells of mixtures with air of lower hydrocarbons ($\Delta = 0.131$ m); R_{PX} — flame acceleration distance (in our case $R_{PX} = 4.9$ m).

When using a ratio that has some theoretical justification [18], we obtain that

$$W_P = U_N\left(R_P/R_K\right)^{3/8} = 15.05\ \text{m/s}, \tag{2}$$

where R_K — critical distance of transition of the laminar combustion to the turbulent one ($R_K = 0.432$ m); R_P — flame acceleration distance (in our case $R_P = 19.5 \cdot 0.95 = 18.525$ m). It must be emphasized that in this situation, the value R_{PX}, used in [16, 17], is different from flame acceleration distance R_P, which is used in [18].

For further calculations, we take the maximum value of the turbulent combustion speed $W_P = 15.05$ m/s.

For conservative calculations, we accept the maximum expert values of the coefficients k_1 and k_2 [16, 17]: $k_1 = 1.8$, which corresponds to the case of initiation of the mixture by turbulent ignition sources, and $k_2 = 4.0$, which corresponds to a case of considerable clutter of space.

As a result, we obtain that the maximum apparent flame speed in our case does not exceed the value $W = 108.4$ m/s.

Using the obtained values of the maximum explosion parameters, the explosion dynamic characteristics were calculated, which can occur during the development of an accident according to the second scenario.

With the accepted parameters of deflagration, the flame front dynamic characteristics will have the form shown in Figure 4. Figure 4 shows the dynamic parameters of the fireball, the dynamics of the apparent flame speed, and the velocity characteristics of the flow accompanying the explosion, which are taken as the initial data: the maximum radius of the fireball will be no more than $R_S = 19.5$ m and the maximum apparent flame speed does not exceed the value $W = 108.4$ m/s.

For these dependences of the flow dynamic parameters, the time dependences of the overpressure were calculated at the points in the space adjacent to the explosion site. The calculations were performed according to the method described in [13] and simulating the formation of an explosive pressure field from an expanding sphere according to the law shown in Figure 4.

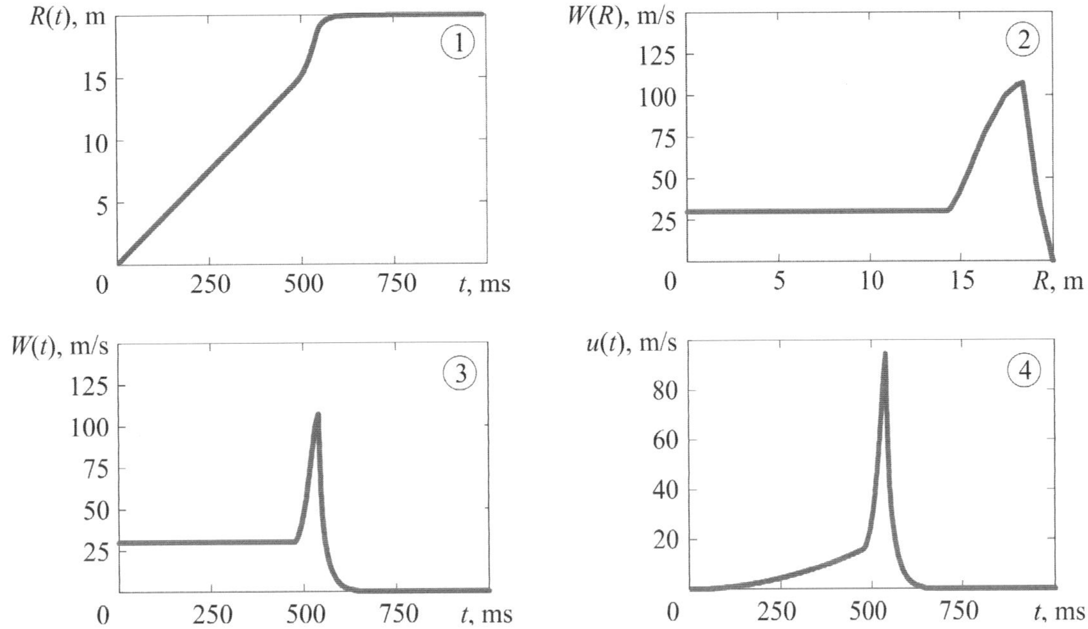

Figure 4. Dynamic dependences of the fireball parameters, the apparent flame speed and the flow velocity at the fireball's boundary: *1* — flame front coordinate versus time characteristic; *2* — apparent flame propagation speed from the flame front position; *3* — dependence of the apparent flame propagation speed versus time parameter; *4* — dependence of the flow velocity at the fireball's boundary versus time parameter.

As a result of the calculations, the field of maximum pressures created by this deflagration explosion was determined [20, 21]. Figure 5 shows isolines of equal maximum pressure, which can emerge in a crash developing according to the second scenario.

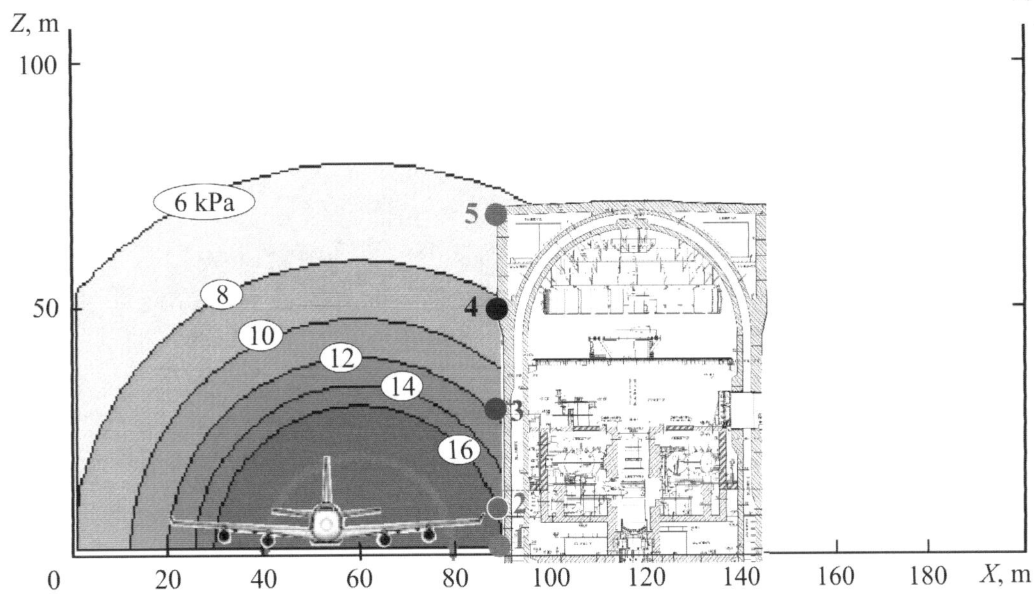

Figure 5. Isolines of equal maximum pressure $\Delta P(X, Z)$ in a crash.

The numerical values of isolines are 6, 8, 10, 12, 14 and 16 kPa. Figure 5 shows the position (highlighted in green) of the fireball at the moment of the explosion completion (the area of the explosion products), the crashing plane, a schematic representation of the standardized power unit building, as well as 5 points on the facade of the power unit building, for which the dynamic parameters of the explosion loading are shown in Figure 6.

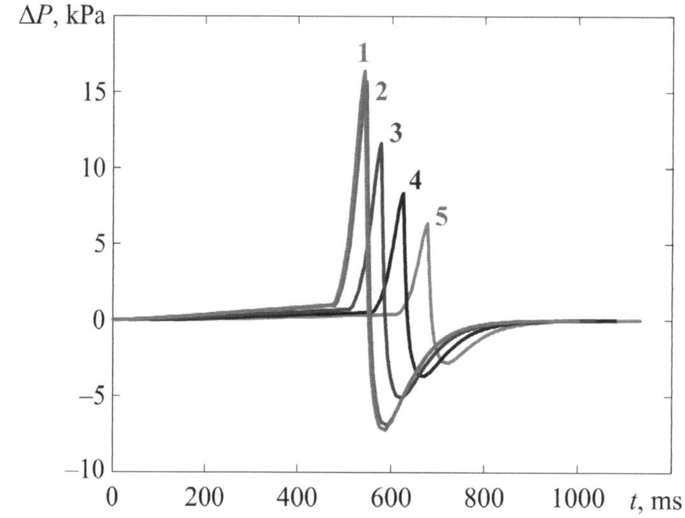

Figure 6. The dynamic parameters of the explosion loading at five points located on the building facade.

The integral parameters of the explosion loading at 5 points of the building facade have the following values (see Table).

Table 1. The integral parameters of explosion loading at 5 points of the building facade.

Parameter name	Explosion loading parameters at points of the facade (see Figure 5)				
	1	2	3	4	5
Maximum explosion pressure ΔP_{max}, kPa	16.4	15.7	11.6	8.4	6.4
Minimum explosion pressure ΔP_{min}, kPa	–7.21	–6.89	–5.10	–3.67	–2.79
Compression phase impulse I, Pa·s	749.9	716.6	530.6	382.2	290.7
Probability of complete destruction V_1, %	4.30	3.60	0	0	0
Probability of medium destruction or loss of life V_2, %	44.4	40.5	18.5	5.26	1.49

The exposure time of the positive phase of the explosion loading, reduced to a triangular shape, will be about 92 ms.

Conditional probabilities of destruction were determined by the methods given in GOST R 12.3.047-2012 OSSS "Fire safety of technological processes".

Figure 7 shows the narrow-band (bandwidth 10 Hz) spectrum of the explosion loading, which is realized at point 1 (the closest point of the building facade to the explosion site, see Figure 5). Figure 7 shows that the vibration loadings are quite high at low frequencies (up to 200 Hz): from 130 to 175 dB. These loadings are quite dangerous for the appliances available in the building. At higher frequencies, sound or acoustic loadings, that is, the noise that witnesses to the crash explosion can hear, will have an intensity of no more than 120 dB, which will be perceived by wider public as a low-frequency hum.

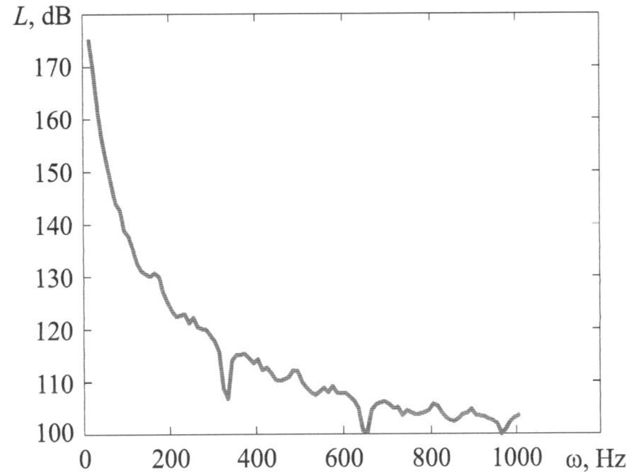

Figure 7. Narrow-band spectrum (10 Hz bandwidth) of explosion loading at point 1, located on the building facade (see Figure 5).

3. Results and discussions

As a result of calculations and based on the correct initial data, the methodology proposed in the article allows:

- obtaining values of the fuel mass capable of forming a highly explosive mixture;
- choosing an appropriate aircraft crash scenario: igneous ball or deflagration explosion of vapor-air mixture;
- determining the time dependence of the fuel vapor concentration in the air;
- evaluating the maximum apparent flame front speed;
- determining the fireball dynamic parameters and the time dependences of the overpressure at points in the space adjacent to the explosion site;
- building up a maximum pressure field created by this deflagration explosion;
- obtaining the integral parameters of the explosion loading: the maximum and minimum overpressure of the explosion, the compression phase pulse, the probability of destruction of buildings and loss of life;
- evaluating vibration loadings from a deflagration explosion.

The calculation results analysis showed that the initial conditions (the presence or absence of a large building near the site where the aircraft crashed; the initial state of the discharge: if the composition consists of vapor fragments or is already well mixed, etc.) have a negligible effect on the initial stage of the highly explosive cloud formation. In all the scenarios of the development of the aircraft crash causing a deflagration explosion at the time of the crash the maximum amount of vapor capable of creating conditions for deflagration does not exceed 7% of the initial vapor discharge mass.

The calculations showed that the proposed methodology allows us to obtain more accurate values of the maximum apparent flame speed, compared with the methods described in the existing Russian regulatory documentation.

From the above calculations (using Boeing-747 as an example), it follows that in the most unfavorable scenario of a plane crash near or directly onto a reactor building, the loadings arising from the deflagration of a fuel-air cloud will have the parameters not exceeding the following values:

- maximum explosion pressure will not exceed $\Delta P_{max} = 16.4$ kPa;
- minimum explosion pressure will not exceed $\Delta P\text{min} = -7.21$ kPa;
- compression phase pulse does not exceed $I = 749.9$ Pa·s;
- maximum probability of complete destruction will not exceed $V_1 = 4.30\%$, and the probability of average destruction of buildings and structures (or loss of life) will not exceed $V_2 = 44.4\%$;
- exposure time of the positive phase of the explosion loading, reduced to a triangular shape, will be not more than 92 ms.

The vibration loadings created by the explosion will have rather high values at low frequencies (up to 200 Hz): from 130 to 175 dB. At higher frequencies, sound or acoustic loadings will have an intensity of not more than 120 dB, which will be perceived by wider public as a low-frequency hum.

The obtained values of explosion loadings arising during any crash of a heavy commercial aircraft are quite severe, but safe for building envelopes of first safety category buildings of nuclear power plants designed for explosion loadings of up to 30 kPa [22].

4. Conclusions

This work presents the most modern methodology for calculating the damage effects of a deflagration explosion associated with crashing an aircraft (A/C). Such calculations are carried out when assessing the consequences of a plane crash near major civil, industrial and energy facilities (such as NPPs).

The article discusses the physical processes that occur during an aircraft crash, which causes the formation of a highly explosive cloud; the methodology for the quantitative assessment of parameters affecting explosion loadings is indicated; the methods of calculating the damage effects of a deflagration explosion during an aircraft crash are presented. When describing the methodology, an example was given of calculating the loadings from a deflagration explosion that occurs when a heavy civilian Boeing-747 crashes onto the NPP site.

The results of calculations carried out according to the methodology described in the article were used to assess the consequences of an aircraft crash on the sites of nuclear power plants in Russia (Novovoronezhskaya, Beloyarskaya, Leningradskaya NPPs), as well as in Hungary (Paks II NPP) and Egypt (El-Dabaa NPP) and were used at the design stage of these facilities. Thus, this methodology has been tested and can be used to calculate the loadings on critically important buildings and structures during a deflagration explosion that occurs when an aircraft crashes.

References

[1] *Structural Analysis and Design of Nuclear Plant Facilities*, editor J. D. Stevenson, New York, American Society of Civil Engineers, pp. 17-25, 1980.
[2] "Report of the ASCE Committee on the Impactive and Impulsive Loads," *Proc. of the Second ASCE Conference "Civil Engineering and Nuclear Power"* vol. V, Tennessee, Knoxville, ASCE, pp. 65-67, 1980.
[3] A. N. Birbraer, and A. J. Roleder, *Extreme Actions on Structures*, Saint Petersburg, Publishing House of the Politechnical University, 2009.
[4] S. E. Bugaenko, *Strength and Reliability of Nuclear Power Plant Designs under Extreme Impacts*, Moscow, Energoatomizdat, 2005.
[5] A. A. Komarov, and N. V. Gromov, "Experimental observation of visible flame propagation rate in accidental deflagration explosions and explosive load reduction," *MATEC Web Conf.*, vol. 251, 02024, 2018.

[6] A. A. Komarov, and N. V. Gromov, "Particular aspects of calculating affecting factors of fireballs emerging from aircraft crash," *MATEC Web Conf.*, vol. 251, 02031, 2018.
[7] A. A. Komarov, N. V. Gromov, and Y. Sokolova, "Assessing destructive capacity of impulse loads," *MATEC Web Conf.,* vol. 86, 01023, 2018.
[8] Y. K. Polandov, and A. D. Korolchenko, "On conditions enabling resonant combustion in gas explosion in non-enclosed volume," *Pozharovzryvobezopasnost/Fire and Explosion Safety,* vol. 27(7-8) pp. 9–23, 2018.
[9] Y. X. Polandov, and A. D. Korolchenko, "About the danger of vibration combustion in gas explosions in the room," *J. Phys.: Conf. Series*, vol. 1425, 012010, 2020.
[10] V. A. Gorev, and D. L. Ovsyannikov, "Evaporation of the liquid methane from the metal surface," *Pozharovzryvobezopasnost/Fire and Explosion Safety*, vol. 28(1) pp. 14–21, 2019.
[11] Y. X. Polandov, and A. D. Korolchenko, "Method for calculating concentration of gas in a non-tight container during gas introduction," *Pozharovzryvobezopasnost/Fire and Explosion Safety,* vol. 27(4) pp 33–41, 2018.
[12] D. A. Korolchenko, A. A. Evich and Y. X. Polandov, "Dynamic Effects at Internal Deflagration Explosions," *IOP Conference Series: Materials Science and Engineering,* vol. 603(5), 052008, 2019.
[13] A. A. Komarov, *Prognosis of loadings of deflagration explosions and assessment of their consequences for buildings and constructions* Dr. Sci. (End.) Diss, Moscow, Moscow State University of Civil Engineering, 2001.
[14] A. A. Abrosimov, and A. A. Komarov, "Mechanisms for the formation of explosive loads in the territory of oil refining complexes," *Oil, Gas and Business*, vol. 50(6), pp. 58–61, 2002.
[15] A. A. Komarov, "Calculation of gas dynamics characteristics of flows during deflagrations at outdoor installations," *Pozharovzryvobezopasnost/Fire and Explosion Safety*, vol. 11(5), pp. 15–18, 2002.
[16] D. Z. Khusnutdinov et al, *Emergency Explosions of Gas Mixtures in the Atmosphere,* Moscow, National Research Moscow State University of Civil Engineering, 2014.
[17] B. S. Rastorguev et al, *Design of Buildings and Structures at Accidental Explosive Influences,* Moscow, Publishing House ASV, 2007.
[18] M. A. Grokhotov, "Calculation of flame propagation rate during deflagration explosion," *Proc. of the 6th International Research and Practice Conference "Firefighting: Challenges, Technologies, Innovation",* Moscow, State Fire Academy of Emercom of Russia, pp 301–303, 2018.
[19] I. Polandov, and D. Korolchenko, "The consideration of the turbulence influence on the gas explosion expansion in non-closed areas," *MATEC Web of Conf.,* vol. 106, 01040, 2017.
[20] A. V. Mishuev, V. V. Kazennov, and A. A. Komarov, 1997, "Some aspects of the protection of the population and territories in case of a special type of emergency (explosions, dam destruction, etc.)," *Protection of the Population and Territories in Emergency Situations in Peacetime and Wartime as an Integral Part of Russia's National Securitys": Abstracts of Reports and Speeche,* Moscow, Vneshtorgizdat, pp. 62–66, 1997.
[21] A. A. Komarov, "Destroy of living buildings at common gas explosions," *Pozharovzryvobez-opasnost/Fire and Explosion Safety,* vol. 13(5) pp, 15–23, 2004.
[22] P. Barr, *Guidelines for the Design and Assessment of Concrete Structures Subjected to Impact,* Report SRD R439 issue 2, Calcheth, United Kingdom Atomic Energy Authority, Safety and Reliability Directorate, 1988.

IOP Conf. Series: Materials Science and Engineering 1066 (2021) 012006 doi:10.1088/1757-899X/1066/1/012006

Strength and Resistance of Structural Steels and Materials

Pavol Juhás [1], Ingrid Juhásová Šenitková [1]

[1] Institute of Technology and Business in České Budějovice, Department of Civil Engineering, Okružní 517/10, 370 01 České Budějovice, Czech Republic

15364@mail.vstecb.cz

Abstract. The paper presents basic information on determining of strength and resistance of structural steels and materials based on test results. The actual European standard EN 1990:2002 and related national standards CSN EN 1990:2004, and other are respected within the paper. The practical application of standard procedures is applied in determining the characteristic and design values of strength and resistance of structural steel S235 used for the production of welded test bars of the previous author's experimental researches. Experimental data on the actual geometric dimensions and material properties of the individual test bars enabled their detailed statistical evaluation. Subsequently, the characteristic and design values of strength and resistance of structural steels used for the production of test bars were determined. The characteristic and design values of strength and resistance determined in this way are compared with the standard ones, which for S235 steel contain valid standards for the design of steel structures.

1. Introduction

The steel structures are designed and made of metallurgical products - materials of suitable standardized shapes and dimensions (bars, sheets, strips), which are rolled from structural steels of various strengths, respectively strength classes. Adequate knowledge of the actual properties, especially the basic strength characteristics of structural steels, is one of the most important prerequisites for reliable and economical design of steel structures.

The basic strength characteristics of structural steels are generally considered to be the yield strength f_y, resp. R_e and tensile strength f_u, resp. R_m from the point of view of practical design. The strength characteristics of structural steels f_y and f_u depend on a number of random variables. For practical purposes, however, they are determined and standardized by concrete values. The determination of the standard values of the yield strength f_y and the tensile strength f_u of the recommended structural steels and materials must therefore have a probabilistic basis [1-3].

The strength of structural steels is influenced by the production itself - chemical composition and technological process. The strength of construction materials is also partially affected by the rolling process. It is therefore not the same, but varies both in length and within the cross-sections of the rolled bars.

The situation can be even more problematic in the case of welded bars consisting of different construction materials from the point of view of strength uniformity. Here, in addition to the already mentioned effects of different melts, rolling, different cross-sections and thicknesses, the strength properties can also significantly affect the welding process. The actual strengths of construction materials are also affected by their dimensions themselves, in particular by the dimensions of their cross-sections, which show certain deviations from the design dimensions. Of course, the actual dimensional deviations of the construction materials can be both positive and negative. Depending on this, dimensional deviations affect the inherent strength and resistance of the construction materials and thus the load-bearing capacity, resp. resistance and reliability of steel bars and structures [4].

From the above it is clear that it is necessary to distinguish between the strength of structural steels and the strength, resp. resistance of structural materials and cross-sections of bars, which also takes into account their possible dimensional deviations. For the design and evaluation of the reliability of the steel structure, the strength of the materials and the material resistance of the bars used for its construction are decisive. The decisive strength characteristic of steel materials is the strength of the structural steels from which they are rolled, therefore the strength of the materials is generally equated with the strength of the structural steels themselves [5, 6].

2. Characteristic and design values steel structures strength and resistance

The strength characteristics of structural steels, in particular their yield strength f_y (R_e) and tensile strength f_u (R_m) are standardized according to individual strength classes on the basis of statistical evaluations of test results. Nominal or characteristic values of f_y and f_u for the individual recommended steels are given by standards for the design of steel structures. Within the European Union for it is mainly EN 1993-1-1:2006 and transformed national standards as are CSN EN 1993-1-1:2005, and other. The values of yield strength f_y and tensile strength f_u are more precisely specified in relevant material standards, as is EN 10025:2004 and transformed national standards CSN EN 10025:2005, and other.

In accordance with the applied method for the design of steel structures according to the limit states for the characteristic values of the yield strength f_y and the tensile strength f_u of structural steels, the failure probability $P_y = 0.05$ is generally accepted. For the calculation and assessment of the load-bearing capacity of steel structures, the design strength values of the structural steels and materials f_d are important. For the design values of strength f_d and resistance f_{Rd}, the failure probability $P_d = P_{Rd} = 0.001$ is generally accepted [1, 2].

The design strength values of structural steels f_d in the actual standards for the design of steel structures are no longer given, but are determined from the respective values of yield strength f_y, or tensile strength f_u and material reliability factors γ_M (γ_{M0}, γ_{M1}, γ_{M2}). Design strength

$$f_{yd} = f_y \,/\, \gamma_M, \text{ consider } \gamma_M = \gamma_{M0} \text{ or } \gamma_{M1} \tag{1}$$

$$f_{ud} = f_u \,/\, \gamma_M, \text{ considered } \gamma_M = \gamma_{M2} \tag{2}$$

Material reliability partial factors γ_{M0}, γ_{M1} and γ_{M2} are standardized. The individual γ_M coefficients according to the CSN 73 1401:1994 are contained in Table 1. The partial factors γ_{M0} and γ_{M1} have different values by strength steel class. The partial factor γ_{M2} by previous standards has a uniform value of 1.3. These values were determined on the basis of a statistical analysis of the actual yield strength and tensile strength of Czech and Slovak structural steels, which were produced in the previous period [6]. Table 1 also shows the material reliability factors γ_{M0}, γ_{M1} and γ_{M2} according to the actual European standard EN 1993-1-1: 2005 and the subsequent transformed standard CSN EN 1993-1-1: 2006.

Table 1: Partial material reliability factors γM.

Carrying capacity	γ_M	**CSN 73 1401:1994 Strength steel classes**			**EN 1993-1-1:2005 CSN EN 1993-1-1:2006**
		S235	**S275**	**S355**	
Cross-sections 1, 2, 3	γ_{M0}	1.1	1.15	1.2	1.0
Cross-section 4	γ_{M1}	1.1	1.15	1.2	1.0
Member´s stability	γ_{M1}	1.1	1.15	1.2	1.0
Weak cross-sections	γ_{M2}		1.3		1.1

The important fact is that in the transformed standard CSN EN 1993-1-1:2006, in accordance with the original European standard EN 1993-1-1:2005 significantly changed values of individual coefficients γ_M were proposed.

The fundamental change is that in the actual standard CSN EN 1993-1-1:2006 the differentiation of γ_M coefficients depending on the strength class of steels is not considered and that the γ_M coefficients have smaller values than in the previous standards CSN 73 1401:1994. These changes are serious in both reliability and economic nature. They should therefore be based on up-to-date and relevant statistical data confirming the higher quality and uniformity of currently produced and recommended structural steels.

Consistently probabilistic-statistical determination of the strength of structural steels and materials requires large sets of necessary data obtained by standardized tests. This presupposes the permanent supplementation and updating of the relevant databases, in accordance with the development of the production of structural steels and the expansion of the material base for the fabrication of steel structures. Such activities are practically not provided in our country. At the same time, they should be carried out at the international level, in accordance with the promoted and applied free market provision of suitable materials that are necessary for the construction of a specific steel structure. It is therefore primarily up to the manufacturers of individual materials and structures to always ensuring their reliability at the required level according to the relevant material standards as well as the applicable standards for the design of steel structures [1, 5].

The following part of the paper presents and analyzes the values of the yield strength of the company, as well as the structural steel S235, which was used for the production of test bars for the research of welded steel structures [4, 5]. To determine the characteristic values of yield strength f_y and design values of strength f_d and resistance f_{Rd} of the materials used, the procedures allowed by the actual European standard EN 1990: 2002 and the transformed standard CSN EN 1990: 2004 are effectively used and practically verified.

3. Evaluation of strength and resistance of structural steels

In the previous period, the author carried out comprehensive experimental programs of research into the elastic-plastic stability and bearing capacity of welded steel bars - beams and columns of materially quasi-homogeneous and combined I cross-sections. The research programs considered included a total of 119 test bars: VP1: 32, VP2: 55 and VP3: 32 [4, 5]. All test bars were produced in the level bridge-production plants under normal production and technological conditions. Prior to the tests themselves, their actual geometric dimensions and material properties were examined in detail. The geometric dimensions of the cross-sections - height h, width b, and flange thickness t_f and web thickness t_w were measured at both ends and in the middle of the bars. The corresponding average values were then considered to be their actual cross-sectional dimensions.

Material characteristics - yield strength f_y, tensile strength f_u, elongation A_5, as well as modulus of elasticity E and modulus of reinforcement E_0 were determined by tensile tests of standardized specimens specially made of flanges and webs of test bars. Seven test specimens were made and tested from each bar, 2 from both flanges and 3 from the web. The corresponding average values were then considered to be the actual material characteristics of the flanges and webs of the individual bars. The actual cross-sectional dimensions and material characteristics were then used to evaluate and analyze the experimental results obtained.

A total of 5 types of Czech-Slovak steels of different strength classes (S235, S355, S440, S530 and S685) were used for the production of all test bars. In accordance with the intention in this paper, only the basic structural steel S235 (11 373.1, 11 375.1), which was used for the flanges and webs of the members of the research program VP1 and for the webs of all members of the programs VP2 and VP3, is evaluated [4, 5]. According to the considered standards CSN 731401: 1994 and CSN EN 1993-1-1: 2006, the nominal and characteristic value of S235 steel is $f_y = 235$ MPa. Partial material factor for this type of steel is $\gamma_{M0} = 1{,}1$, resp. $\gamma_{M0} = 1{,}0$. Therefor standard design strengths according these standards $f_d = 213{,}64$ MPa resp. 235 Mpa.

The determined yield strengths of flanges f_{yf}, webs f_{yw} and flanges and webs f_y as well as cross-sectional deviations of flanges ϕ_{Af}, webs ϕ_{Aw} and cross-sections ϕ_A of the test bars were in detail evaluated. For illustration see Figure 1. Statistical characteristics of the relative values of cross-sectional areas of flanges ϕ_{Af}, webs ϕ_{Aw} and cross-sections ϕ_A of test bars made of steel S235 (research programs VP1, VP2 and VP3) are presented in Table 2.

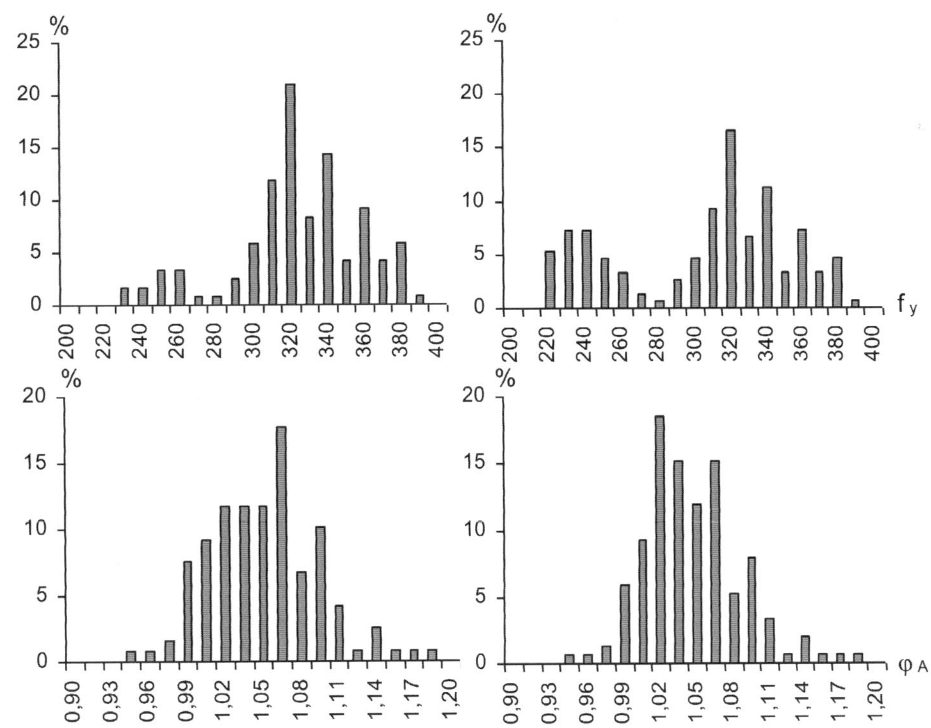

Figure 1. Histograms of the yield strength f_{yw}, f_y and relative values φ_{Aw}, φ_A of the cross-sections of test bars made of steel S235.

Table 2. Statistical characteristics of the relative values of cross-sectional areas of flanges φ_{Af}, webs φ_{Aw} and cross-sections φ_A of test bars made of steel S235.

	VP1			VP2	VP3	VP1+VP2+VP3	
	φ_{Af}	φ_{Aw}	φ_A	φ_{Aw}	φ_{Aw}	φ_{Aw}	φ_A
n_a	32	32	32	55	32	119	119
m_a	1.0207	1.0538	1.0344	1.0343	1.0481	1.0433	1.0433
s_a	0.0160	0.0421	0.0228	0.0488	0.0347	0.0441	0.0441
γ_a	0.0157	0.0400	0.0220	0.0472	0.0331	0.0423	0.0423
a_a	0.5450	1.1708	0.6367	0.6133	-0.4612	0.4542	0.4542
$\varphi_{a,min}$	0.9925	0.9929	0.9948	0.9415	0.9886	0.9414	0.9414
$\varphi_{a,max}$	1.0595	1.1791	1.0808	1.1700	1.0898	1.1791	1.1791
$\Delta\varphi_a$	0.0670	0.1862	0.0859	0.2285	0.1012	0.2376	0.2376

The individual statistical quantities in the Table 2 mean: n_a - number, m_a - average value, s_a - standard deviation, v_a – coefficient of variation, a_a - slant, $\varphi_{a,\,min}$ - minimum value, $\varphi_{a,\,max}$ - maximum value, $\Delta_{\varphi a}$ - difference of extreme values.

Using the relevant statistical characteristics and applying the corresponding standard relationships, the characteristic values of yield strength f_y and design strength f_d and resistance f_{Rd} for the flanges and webs of the test bars were determined within the individual research programs and summaries. In the case of design resistances f_{Rd}, the determined dimensional, resp. cross-sectional deviations not taken into account in the standards considered. The determined values of f_y, f_d and f_{Rd} are summarized in the following Table 3 (assumed logarithmic-normal distribution of occurrence).

The considered standards EN 1990: 2002 and CSN EN 1990: 2004 for determining the characteristic and design strength values of structural steels by means of tests consider, depending on the existing results and knowledge, the following two cases a, b.

Table 3. Characteristic and design values of strength and resistance of structural steel S235.

Strength/resistance		Program			
		VP1	VP2	VP3	VP1+VP2+VP3
f_{yf}	a	205.840			205.840
	b	206.618			206.618
f_{yw}	a	228.781	290.748	293.425	265.852
	b	230.564	291.558	294.188	266.290
f_y	a	198.272			230.415
	b	199.181			230.853
f_{df}	a	183.743			183.743
	b	187.358			187.358
f_{dw}	a	181.098	253.971	271.351	225.414
	b	188.509	257.646	275.017	227.340
f_d	a	153.550			181.396
	b	157.221			183.184
$f_{Rd,f}$	a	186.380			186.380
	b	190.148			190.148
$f_{Rd,w}$	a	187.188	252.357	271.351	229.318
	b	195.169	256.538	275.017	231.421
f_{Rd}	a	158.073			185.548
	b	161.888			187.432

a) available only the results of small number selection tests n
b) previous knowledge providing a priori results is also available in addition to the results of sample tests

The first case (a) is general and unambiguous if only the results of a limited number of n sample tests are available and no previous knowledge and necessary statistical characteristics are available (mean value my, standard deviation s_y and coefficient of variation v_y). In EN 1990: 2002 and CSN EN 1990: 2004, it is recommended to consider the second case (b) with a conservative estimate of the coefficient of variation v_y.

However, in the context of the above remarks on the nature and relationship of previous statistics to the new set of material test results to be used for a particular design, any estimate of the coefficient of variation v_y seems problematic. With thorough and sufficient verification of all materials used for a particular construction, the corresponding coefficient of variation v_y should probably be considered representative and known. If the first method (a) is used, then the coefficient of variation v_y should not be considered to be less than 0,1 according to these standards. With regard to the number of tests, it is based on the assumption of a normal - symmetrical distribution or a two-parameter logarithmic-normal distribution of the occurrence of the evaluated strength. The corresponding inclination a_y is not considered, but it can be very significant.

The obtained and partial presented histograms and statistical characteristics, as well as the calculated characteristic and design values of strength and resistance of the evaluated structural steels and materials of the considered test bars allow discussion and evaluation of the results from several relevant perspectives.

4. Conclusions

The examined values of yield strength $f_{yf,i}$ and $f_{yw,i}$ can be responsibly considered in accordance with [4, 5] when assessing the load-bearing capacity of individual test bars. For test bars of individual research programs, reliably consider also the minimum values of the determined yield strength $f_{yf,min}$ and $f_{yw,min}$. These values are close to or higher, up to significantly higher than the corresponding standard design strength value $f_{d,CSN}$.

The test bars were produced in two stages with a time lag at two manufacturers, the values found for the yield strength of flanges and webs f_{yf} and f_{yw} are probably relatively different, however, the effect of thicknesses and rolling of the materials used is also evident. The yield strength of f_{yf} flanges is generally smaller than the yield strength of f_{yw} webs. However, even the minimum value of $f_{yf,min}$ is practically equal to the design value of strength $f_{d,CSN}$ determined with the coefficient $\gamma_{M0} = 1.1$. Certain differences were also found in the proposed and actual cross-sectional areas of flanges and webs of test bars of the individual research programs, they are characterized by the detected deviations ϕ_a. Due to their size ($\phi_{a,min} = 0.9414$ and $\phi_{a,max} = 1.1791$), the test members considered were also taken into account when determining the design values of the material resistance f_{Rd}.

From the point of view of the paper and the adequacy of determining the characteristic and design values of strength and resistance of structural materials based on test results, it is important to compare the determined characteristic value of yield strength f_y with the design strength f_d and resistance f_{Rd} - Table 3.

Determination and verification of strength characteristics of structural steels and materials on the basis of relevant test results and presented procedures according to the actual standards is justified and necessary especially for generally demanding structures. The results of the presented evaluation encourage caution in the application of partial reliability coefficients of structural steels and materials according to the actual standards for the design of steel structures.

References

[1] A. Mrázik, "Theory of reliability of steel structures," VEDA, Bratislava 1987.
[2] P. Juhás, et al., "Design of steel structures," *Commentary on STN 73 1401:1998*. Slovak Institute for Standardization, Bratislava, Slovakia, 2001.

[3] V. Raizer, *Reliability of structures – analysis and application*, Backbone Publishing Co., New York, 2009.
[4] P. Juhás, *Quality and geometric deviations of bars and cross-sections of steel structures*, TU Košice, ELFA, Košice, Slovaka, 2007.
[5] P. Juhás, *Reliability and strength of materials of steel* structures, TU Košice, ELFA, Košice, Slovakia, 2009.
[6] A. Mrázik, Z. Sadovský, and J. Kriváček, "Catalog of statistical data on yield strength, tensile strength and elongation of steels from the previous period (four parts)," ÚSTARCH SAV, Bratislava, Slovakia 1992.

IOP Conf. Series: Materials Science and Engineering 1066 (2021) 012007 doi:10.1088/1757-899X/1066/1/012007

Properties of Selected Polyurethane Sealants in the Sealing of Cement-based Materials

Adam Boháček [1]

[1]Brno University of Technology, Faculty of Civil Engineering, Veveří 331/95, Brno 602 00, Czech Republic

bohacek.a@fce.vutbr.cz

Abstract: Since the rise of cement-based building elements, there has been a need to seal these materials. Therefore, this article deals with the sealing of cement-based materials, specifically glass cement. The aim of this work is to find a sealant for sealing this problematic substrate, to verify its suitability and to show that polyurethane sealants are also important in the construction market. The problem of a cement-based substrate lies mainly in the dustiness of its surface, where there are always small dust particles on its surface. The main method for testing these sealants is the tensile test, which checks both the properties of the sealant itself and the contact joint between the sealant and the base material. Tensile test specimens are made of glass-cement plates and three representatives of one-component flexible polyurethane sealants are selected as the sealing material. From the results of the tensile test, it can be concluded that, with the appropriate choice of sealant and primer, problematic cement-based materials with good results can be sealed, provided that the grinding is carried out correctly.

1. Introduction

Our society has been cementing materials practically since the beginning of construction history. We still need to bond various materials together today, and we can see cemented joints virtually everywhere. Naturally, cemented joints have developed over time, but now we have shifted from the original natural materials to the area of the chemical industry. Due to the significant expansion of the chemical industry, great advances have been made in the development and application of sealants in the field of cemented joints over the last two decades [1].

Cement joints are now a common part of the building industry, and we could not imagine modern construction without them. Their main aim is to prevent substance exchange between the construction materials and external environment. They also ensure thermal and sound insulation, and some even contribute to the construction's fire resilience [2]. We can find them virtually in all parts of building constructions, for example between prefabricated concrete elements in façades, around windows and doors, in the joints between floors and walls etc. Even though we can find cemented joints in various places, it is evident that different demands are imposed on different joints. So obviously there are significant differences between the individual cemented joints [2,3].

And as a result of the different demands on cemented joints, when designing a cemented joint, we have to take into account all the variables which might negatively influence the cemented joint. To the greatest extent these variables include external climatic influences [3]. If we assume that a cemented joint will be exposed to external climatic influences, it is necessary to select a sealant where the manufacturer states that it is resilient to negative climatic phenomena [4-6]. On the Czech market, there is really a huge number of different sealants from different manufacturers, which differ in terms of quality and price. This can make it difficult for the regular user to navigate such an extensive offer and choose the right sealant. And this situation is also complicated by the fact that the manufacturer's recommendations concerning resilience of the sealant do not always correspond with reality [7-10] This is another reason why the following research was undertaken, which is intended to compare the regular sealants available on the Czech market [11-13] and their properties after exposure to negative climatic influences.

2. Material

For the purposes of this research, it is necessary to select the base material, sealant recommended by the manufacturer for the needs of cementing the selected base material, and a suitable primer.

The decisive factor for choice of the material used in this survey is how difficult the material is for the creation of cemented joints and its availability on the Czech market. Glass reinforced concrete was selected as the base material based on these criteria. This material has a large number of dust particles on its surface which greatly weaken the cohesion between the sealant and its surface layer. This property should be in part improved by the application of a primer before the actual sealant.

As far as the actual sealants are concerned, consideration is given to the availability for the regular user and the optimum price offer as well as recommendation of the manufacturer that the sealant is suitable to use for cementing the base material chosen by the author. Following a careful survey of the Czech market, three polyurethane sealants from three different manufacturers in various price categories were chosen. Three primers were then selected for these sealants. These primers were selected at the recommendation of the manufacturer.

Selected sealants (Table 1) and primers (Table 2) are further specified in the tables below.

Table 1. Properties of sealants.

Properties	Sealant A	Sealant B	Sealant C
Density [g/cm^3]	1.23	1.3	1.39
Tensile strength [N/mm^2]	1.4	1.5	3.8
Consistency	Thixotropic paste	Thixotropic paste	Thixotropic paste
Elastic modulus of elasticity [N/mm^2]	0.7	0.6	0.7
Hardness according to Shore A	40	37	40

Table 2. Properties of primers.

Properties	Primer A	Primer B	Primer C
Density [g/cm^3]	1.01	1.0	1.05
Consistency	Low viscosity liquid	Liquid	Liquid
Dry matter content [%]	-	34	70

3. Methodology

The test methods described in this article are based on the valid Czech technical standard ČSN EN ISO 8340 Building construction - Sealants - Determination of tensile properties at maintained extension [14]. This standard gives a precise definition of the test body and a precise tensile test procedure. But it does not precisely define the test rig.

The test body consists of two base boards of precisely defined dimensions, but there can be changes as long as the dimension of sealant and adhesion area remain the same. It also consists of two spacers and the actual sealant. A test body was designed consisting of 50 x 30 mm base boards where the thickness of the base boards was 12.5 mm. In order to ensure that the cemented joint is properly exposed to the air, the spacers are made of wood, are 50 mm high and have plan dimensions of 12 x 12 mm. The base boards (along with two wooden spacers) define the boundary of the applied sealant and together make up these individual parts of the test body. The standard also gives the minimum number of test bodies for one test (three test bodies). But for the purposes of this test, five test bodies were used in order to provide a better evaluation of the test results. The test bodies were constructed at a constant temperature of sealant and test boards (23 ± 2) °C, and the instructions of the sealant manufacturers were adhered to. These may include a condition to use a base coat, so-called primer. And according to the standard it is necessary to adhere to the designated conditions: to ensure that there are no air bubbles, to press the sealant onto the contact areas of the base bodies, and to smooth the surface of the sealant to the level of the test bodies and spacers.

Before the actual test it is necessary to cure the test bodies, i.e., to store them according to the procedure defined by the Czech standard. The test bodies are first of all left at a temperature of (23 ± 2) °C and relative humidity of (50 ± 5) % for 28 days so that the applied sealant can cure properly, and then they undergo the following storage cycles three times:

1. 3 days in drying room at the temperature of (70 ± 2) °C
2. 1 day in distilled water at the temperature of (23 ± 2) °C
3. 2 days in drying room at the temperature of (70 ± 2) °C
4. 1 day in distilled water at the temperature of (23 ± 2) °C

The standard defines the procedure of the test for determining tensile properties at maintained extension used for the given survey, but as already stated it does not define the actual test rig to be used. For this reason the test rig shown in figure 1 was built as part of the preceding research into the sealants. This tensile test is performed in two variants - at the temperature of (23 ± 2) °C and (-20 ± 2) °C. The principle of the test is that the test bodies which have undergone the prescribed cycles are exposed to the prescribed temperature and then stretched at a rate of (5.5 ± 0.7) mm/min to 25 % more than their original length.

The first variant is the performance of a tensile test at temperature of (23 ± 2) °C where the test body is stored for 24 hours at the temperature of (23 ± 2) °C and relative humidity (50 ± 5) % after cycling. After this period the test bodies are put in the test equipment shown in figure 2 and stretched at a rate of (5.5 ± 0.7) mm/min to 25 % more than their original length at a temperature of (23 ± 2) °C. This extension is maintained for 24 hours at a temperature of (23 ± 2) °C. After this there is a visual inspection of defects in adhesion and cohesion, which are then measured with a calliper.

The second variant is performance of a tensile test at the temperature of (-20 ± 2) °C where the test bodies are kept at the temperature of (-20 ± 2) °C for at least 4 hours. The following procedure is similar to the first variant. The test bodies are put in the test device and stretched at a rate of (5.5 ± 0.7) mm/min at a temperature of (-20 ± 2) °C to 25 % more than their original length. This extension is maintained at a temperature of (-20 ± 2) °C for 24 hours. Then defects in adhesion and cohesion are identified and measured with a calliper. In this case the defects are identified after the test bodies have been taken out of the cooling chamber and thawed out [14].

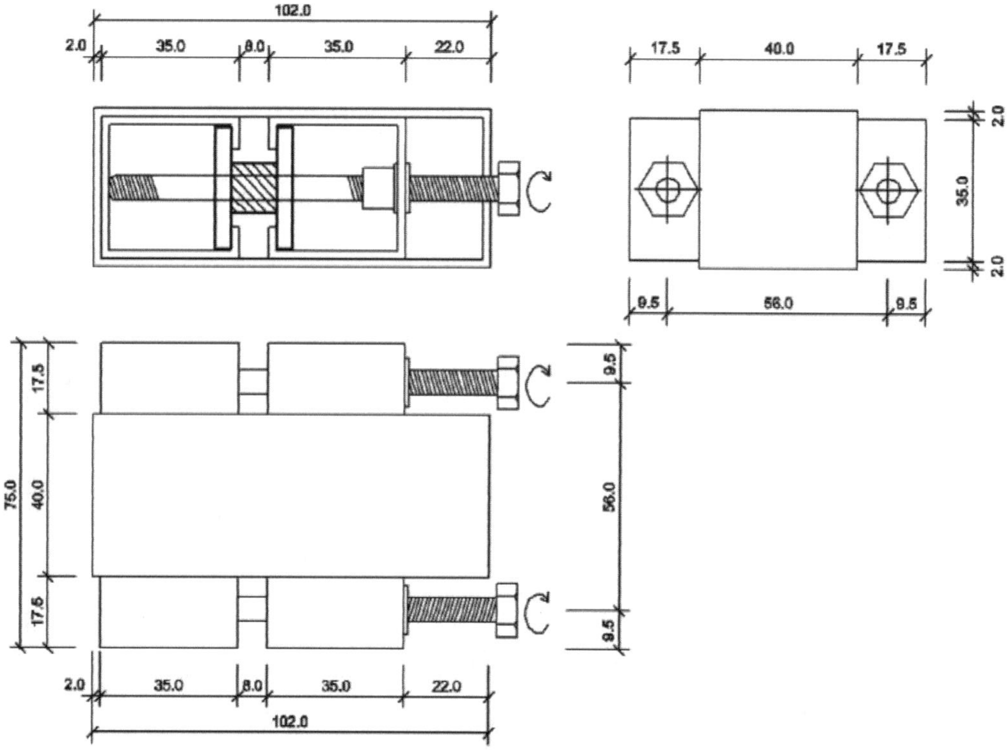

Figure 1. Test rig for maintained extension.

4. Results

Tests were performed on five test samples of each sealant. The evaluation of the test results was performed by a visual inspection of the test sample and then measuring using a calliper. For greater transparency of result evaluation, the testing is divided into two sections - testing at the temperature of (23 ± 2) °C, the results of which are given in table 3, and testing at the temperature of (-22 ± 2) °C, the test results of which are recorded in table 4. Both tables contain an evaluation of the joint between the sealant and the base material, and any breakage is described.

A joint evaluated in the tables as Failed following the visual inspection and calliper measurement displays signs of breakage. Such a joint is shown for testing at temperatures (23 ± 2) °C in figure 2, and for testing at temperatures (-22 ± 2) °C in figure 4.

In contrast a joint evaluated as Did not fail after inspection does not display signs of breakage. For the purposes of clarity, an unbroken joint after testing at the temperature (23 ± 2) °C is shown in figure 3, and joint after testing at temperature of (-22 ± 2) °C in figure 5.

If the tested samples display breakage, the sealant has separated from the base material on one side.

As you can see from the table of results, it does not contain test results for the third sealant. It was not possible to test and measure this sealant because all the samples disintegrated during cycling.

Table 3. Test according to ČSN EN ISO 8340 at temperature of (23 ± 2) °C.

Type of sealant	Polyurethane sealant					
	Sealant A			Sealant B		
Base material	No	Joint	Breakage	No	Joint	Breakage
Glass reinforced concrete	1	Failed	one-sided separation when stretched by 25 %	1	Failed	one-sided separation when stretched by 25 %
	2	Failed	one-sided separation when stretched by 25 %	2	Failed	one-sided separation when stretched by 25 %
	3	Failed	one-sided separation when stretched by 25 %	3	Did not fail	none
	4	Failed	one-sided separation when stretched by 25 %	4	Did not fail	none
	5	Failed	one-sided separation when stretched by 25 %	5	Did not fail	none

Figure 2. Sealant A, sample No 3.

Figure 3. Sealant B, sample No 4.

Table 4. Test according to ČSN EN ISO 8340 at temperature of (-22 ± 2) °C.

Type of sealant	Polyurethane sealant					
	Sealant A			Sealant B		
Base material	No	Joint	Breakage	No	Joint	Breakage
Glass reinforced concrete	1	Failed	one-sided separation when stretched by 25 %	1	Failed	one-sided separation when stretched by 25 %
	2	Failed	one-sided separation when stretched by 25 %	2	Did not fail	none
	3	Failed	one-sided separation when stretched by 25 %	3	Did not fail	none
	4	Failed	one-sided separation when stretched by 25 %	4	Failed	one-sided separation when stretched by 25 %
	5	Failed	one-sided separation when stretched by 25 %	5	Did not fail	none

Figure 4. Sealant A, sample No 4.

Figure 5. Sealant B, sample No 2.

5. Analysis of results

As can be seen from the above table of results, during the testing of samples at the temperature of (23 ± 2) °C and temperature of (-22 ± 2) °C the test samples displayed only one defect - a one-sided separation of the sealant from the base material. There were varying values for this separation, but even at low values such a sealant must be evaluated as non-compliant.

In the case of testing at the temperature of (23 ± 2) °C we can see in the table of results that sealant A displays far worse cohesion between the sealant and base material than sealant B. In the case of the first of the sealants, all the tested samples displayed signs of breakage.

There is also clear breakage of the cemented joints in all the tested samples of sealant A at the temperature of (-22 ± 20) °C. Thus, in general terms sealant B displays better results in this specific testing than sealant A.

6. Conclusion

It is evident from the author's measurement that each of the tested sealants displays shortcomings when used for cementing the selected base material. As a result it must be stated that in spite of the recommendations given by the manufacturer and specified suitability of the sealant for cementing glass reinforced concrete base materials, none of the tested sealants can really be recommended as suitable for cementing.

Acknowledgements

The article was written with the support of the junior specific research of the Brno University of Technology, Faculty of Civil Engineering, managed under the designation FAST-J-20-6359.

References

[1] E. Petrie, *Handbook of adhesives and sealants*, New York: McGraw-Hill company, 2009.
[2] J. Klosowski, and A. T. Wolf, *Sealants in construction*, 2nd edition. Boca Raton: Taylorand Francis Group, 2016.
[3] K. L. Mittal, and A. Pizzi, *Handbook of sealant technology*, Boca Raton: Taylor and Francis Group, 2009.
[4] M. Z. L. Chew, X. Zhou, and S. T. Tan, "On-site weathering of sealants under tropical conditions," *Construction and Building Materials*, vol. 18(4), pp. 287-293, 2014.
[5] H. Stögbauer, and A. T. Wolf, "The influence of heat ageing on one-part construction silicone sealants," *Construction and Buildings Materials*, vol. 5(1), pp. 27-32, 1991.
[6] A. T. Wolf, "Movement capability of sealants," *Construction and Building Materials*, vol. 5(1), pp.127 – 134, 1991.

[7] B. Nečasová, P. Liška, and J. Šlanhof, "Performance of selected polyurethane joint sealants in concrete structures," *MATEC Web of Conferences*, vol. 146, 02015, 2018.
[8] J. Šlanhof, P. Liška, and B. Nečasová, "Evaluation of test methods for testing of sealants, Advances and Trends in Engineering Sciences and Technologies II," *Proceedings of the 2nd International Conference on Engineering Sciences and Technologies, ESaT 2016*, pp. 643-648, 2017.
[9] J. Šlanhof, M. Šimácková, P. Liška, and B. Nečasová, "Sealing possibilities of glass fibre reinforced concrete elements and fibre - Cement facade cladding panels," *Proceedings of an International Scientific Conference in Bratislava - Construction Technology and Management CTM 2014*, pp. 447-455, 2014.
[10] P. Liška, B. Nečasová, J. Šlanhof, and M. Šimáčková, "Determination of Tensile Properties of Selected Building Sealants in Combination with High-pressure Compact Laminate (HPL)," *Procedia Engineering*, vol. 108, pp. 199-205, 2015.
[11] Den Braven. Accesible from: https://www.denbraven.cz/
[12] Sika. Accesible from: https://cze.sika.com
[13] Mapei - adhesives, sealants, chemical products for building. Accesible from: https://www.mapei.com
[14] ČSN EN ISO 8340 Building construction - Jointing products - Sealants - Determination of tensile properties at sustained elongation Prague: Czech Standards Institute, 2006.

Mixed Low-density Demolition Waste in Production of Lightweight Cement-based Composites

Jaroslav Pokorný[1], Jiří Šál[1], and Radek Ševčík[1, 2]

[1] Faculty of Technology, Institute of Technology and Business, Okružní 517/10, 370 01 České Budějovice, Czech Republic

[2] Institute of Theoretical and Applied Mechanics of the Czech Academy of Science, Prosecká 809/76, 190 00 Prague, Czech Republic

jaroslav.pokorny@mail.vstecb.cz

Abstract. Lightweight concretes are traditionally used composites intended for non-bearing applications. In their composition, various types of lightweight natural as well as industrially produced aggregates are employed. Regard to the considerable aggregates consumption, the potential application of mixed low-density demolition waste (MDW) in production of cement-based composites was studied in this paper. In mixes with maintained workability, silica sand was replaced by MDW in the amount of 0 – 100 vol. %. On hardened developed composites, basic physical properties, strength properties and dynamic moduli were determined and compared with control material after 28 days of water curing. Due to reduced weight of waste aggregate compared with the silica sand, hardened composites showed importantly lower bulk density. However, their increasing porosity induced by higher MDW additions led to the decrease of compressive and flexural strengths and dynamic moduli values. With respect to the maintaining of mechanical resistance, quantities of incorporated MDW not exceeding 25 % were chosen as an optimal replacements ratio.

1. Introduction

Due to the constant economic growth of last years, the construction industry is still developing area consuming a considerable amount of natural non-renewable resources [1]. Its estimated that about 50 % of extracted nature commodities is utilized for this purpose. With nature resources depletion, the governments all over the world are driven to support environmental policies and look for alternative materials [2]. New developer projects as well as increasing urbanization connected with meeting of human needs bring additional impacts on the environment. Accordingly, over 859.5 million tons of construction and demolition waste (CDW) is generated across the EU annually [3]. In view to the European directives, produced CDW cannot be disposed in landfills. The major CDW amounts should be recycled and reused [4]. For this reason, recycled concrete and ceramic aggregates found the applications as subbase layers or filling material of low strength composites. CDW often includes other materials, such as packaging materials, plastics, insulating materials, etc. whose source separation is necessary to maintain their utility value [5]. Nevertheless, in some cases the separation process can be problematic or highly financially demanding and thus different blends with inferior properties are formed.

With the aim to reduce the amount of unseparated waste from construction industry, this work is focused on the potential application of mixed low-density demolition waste (MDW) in the manufacturing of cement-based composites. The results pointed out to the possibility of the production of environmentally friendly lightening materials.

2. Experimental

2.1. Used materials

In composites production, CEM I 42.5 R (Českomoravský cement, Ltd., Czech Republic), meeting the specifications given in the EN 197-1 [6] standard, was applied. Its oxides composition and basic material properties, available from the producer, are summarized in Table 1 and 2. Natural silica sand (Filtrační písky, Ltd., Czech Republic), in the fraction 0.0/2.0 mm, was used as reference aggregate. This sislica sand was substituted by mixed low-density demolition waste (MDW) in the amount from 0 % up to 100 vol. %. The material blend was consisted from cotton-polyester mix, paper-based packaging materials, polyethylene and cellulose in the concentrations of 56.0 wt. %, 24.4 wt. %, 12.2 wt.% and 7.1 wt. % respectively. Moreover, polyether urethane was found in the concentration less than 1 wt.%. The particle size distribution of both sand and its preplacing material are introduced in Table 3. From indicated data, higher occurrence of rougher MDW particles in the interval from 1.0 mm to 2.0 mm are visible.

Table 1. Oxides composition of CEM I.

Oxide composition (wt.%)	
SiO_2	19.62
Al_2O_3	4.80
Fe_2O_3	3.34
CaO	63.70
MgO	1.42
K_2O	0.75
Na_2O	0.19
SO_3	3.10
Cl^-	0.04

Table 2. Material properties of applied binder.

Powder density ($kg \cdot m^{-3}$)	Specific density ($kg \cdot m^{-3}$)	Specific surface area ($m^2 \cdot kg^{-1}$)	Loss on ignition at 1000°C (wt.%)
980	3 110	408	3.42

Table 3. Particle size distribution of sand and replacing aggregate.

Material	d_{10}	d_{50} (mm)	d_{90}
Sand	0.20	0.72	1.67
MDW	0.17	1.05	1.80

2.2. Mixes design and sampling

The overview of compositions of prepared composites is listed in Table 4. Control mix was comprised from 487.8 kg of Portland cement per 1 m^3 and three times higher amount of silica sand. In the case of other four mixes, sand was replaced by MDW in the quantity up to 100 % by volume with the step of 25 %. Silica sand replacing material disposed a considerable water absorption of 360 % (about 0.85 % measured for sand) and thus was left to soak with water in the plastic container for 24 hours before the application. In this context, the amount of added water was different in order to kept the constant value of spreading of 160 × 160 ± 10 mm (EN 1015-3 [7]) for all fresh composites.

From fresh mixes, prismatic samples with dimensions of 40 × 40 × 160 mm and 100 mm cubes were casted and compacted with the help of a vibrating table (Matest, S.p.A, Italy) for 60 s. Placed tested specimens were covered by plastic foil and left under the laboratory conditions at 20 ± 1 °C and 45 ± 5 % of relative humidity for 24 hours. Subsequently, demoulded hardened specimens were cured under water at conditions mentioned above for other 27 days.

Table 4. Compositions of developed mixes.

Mix	Content (kg·m^{-3})			
	CEM I	Sand	MDW	Water
Control	487.8	1 464.4	-	243.9
MDW-C 25	490.5	1 103.7	37.2	147.2
MDW-C 50	493.1	739.8	74.8	98.6
MDW-C 75	495.8	372.0	112.8	49.6
MDW-C 100	498.4	-	151.1	-

2.3. Measured properties

2.3.1. Particle size distribution. Particle size distribution (PSD) of aggregates was measured with the usage of sieving analysis according to the standard EN 933-1 [8] on dried aggregates samples. Each sample was sieved through the set of sieves having mesh size of 2.0; 1.0; 0.5; 0.25; 0.125 and 0.063 mm and shaken by vibratory apparatus Retsch AS 200 (Germany).

2.3.2. Basic physical properties. Basic physical properties measured after 28 days on hardened samples, bulk density, saturated water content and open porosity under vacuum conditions, were determined. Bulk density measurements were performed following procedure described in the standard EN 1015-10 [9]. The relative expanded uncertainty of the density tests was 2.7 %. The saturated water content and further open porosity were measured with the gravimetric method according to the EN 1936 [10]. In both cases, the expended uncertainty did not exceed 5 %. Moreover, powder densities of aggregates were measured in the agreement with the EN 1097-3 [11].

2.3.3. Strength parameters and dynamic moduli measurements. Flexural and compressive strengths tests were performed with the usage of hydraulic press Servo Plus Evolution (Matest, S.p.A., Italy) with disposing loading capacity of 300/15 kN accordingly to the EN 1015 – 11 [12]. Prismatic samples with dimensions of 40 × 40 × 160 mm in three-point bending setting were tested. Further, maximum compressive forces were recorded on the broken halves of prisms [13]. The expanded measuring uncertainty of both mentioned procedures was 2.5 %. Moreover, dynamic moduli measurement with the device Dio 562 NLF (Starmans Electronics) was carried out. The relative expanded uncertainty of the dynamic moduli measurements of 2.4 % was recorded.

3. Results and discussion

Powder density measurements performed on both aggregate samples pointed out to importantly lower value of 150 kg·m^{-3} attributed to artificial aggregate. On the other hand, used common silica sand had approximately tenfold higher powder density (1 492 kg·m^{-3}, detected in loose state). Significantly different values of powder density thus influenced the obtained basic physical properties given in Table 5. At first sight, the gradual decrease in bulk density caused by the increased incorporation of MDW is visible. Already at MDW addition of 25 vol. %, the bulk density of hardened specimens was dropped under 1 800 kg·m^{-3} which is the limiting value for lightweight concretes according to the EN 206-1 [14]. On contrary, open porosity was detected to be considerably higher for MDW composites. Control samples average values of absorbed water was detected to be 189 kg·m^{-3} and these samples exhibited average value open porosity of 19.8 %. With the substitution of natural silica sand with MDW, saturated water content and open porosity started to increase. Fully substituted composite, MDW-C 100, showed more than three times higher values of these physical quantitates in the comparison with the control sample. In this sense, the application of MDW promoted an important weight reduction of produced composites bringing benefits like decrease of transportation costs of precast products, reduction of structural dead loads, etc.

Table 5. Basic physical properties measured on hardened composites.

Mix	Bulk density ($kg \cdot m^{-3}$)	Saturated water content ($kg \cdot m^{-3}$)	Open porosity (%)
Control	2 055	189	19.8
MDW-C 25	1 723	295	30.9
MDW-C 50	1 424	407	40.5
MDW-C 75	1 162	480	48.4
MDW-C 100	760	625	62.7

Mechanical properties, flexural strength, compressive strength and dynamic moduli values, obtained on tested samples are summarized in Table 6. As a consequence of growing open porosity of waste aggregate dosed composites, the continuous decrease of strength performance was recorded. Flexural strengths values of MDW-C 25, MDW-C 50, MDW-C 75 and MDW-C 100 were detected to be lower for 19 %, 45 %, 60 % and 78 %, respectively, in the comparison with control sample. Likewise, the considerable decease compressive strengths values were calculated. In the comparison with control sample, values of compressive strength were detected to be lower for 47 %, 79 %, 88 % and 95 % of MDW-C 25, MDW-C 50, MDW-C 75 and MDW-C 100, respectively. From reported values, it is visible that the decrease of compressive strength was more noticeable. Less steep decrease of flexural strength values could be ascribed to the partial reinforcing capability of cellulose and cotton fibres occurring in the artificial aggregate blend as reported e.g. in [15]. Similarly, with higher representation of MDW in produced composites, the dynamic moduli values dropped dramatically. The observed trend is in line with previous research done by Malešev et al. [16] in which it was described that the usage of lower density aggregates resulted in decrease of dynamic moduli. From quantitative point of view, all MDW applications exceeding 25 vol. % adversely influenced mechanical properties.

Table 6. Mechanical properties measured on 28 days cured samples.

Mix	Flexural strength (MPa)	Compressive strength (MPa)	Dynamic moduli (GPa)
Control	8.0	56.6	35.6
MDW-C 25	6.5	29.9	18.6
MDW-C 50	4.4	11.9	8.0
MDW-C 75	3.2	7.0	4.7
MDW-C 100	1.8	3.0	2.0

4. Conclusions

The influence of low-density demolition waste from construction industry as a potential alternative of silica sand in manufacturing of lightweight cement-based composites was studied. Applied replacing aggregate had importantly reduced powder density compared to silica sand and thus considerable lightening of produced composites was achieved even for samples with the lowest of sand replacement. However, high rate of open porosity up to 62.7 %, attributed to MDW modified samples, resulted in important strengths losses. The compressive strength measured for MDW-C 100 was reduces of about 95 % with respect to the control material. In view of examined properties and to maintain acceptable strength performance of produced composites, an optimal amount of MDW seems to fall in the range from 0 to 25 vol. %. Nevertheless, the increasing open porosity of modified tested samples may improve thermal performance of designed composites. This will be further investigated in the subsequent research.

Acknowledgments
The authors gratefully acknowledge the funding received from the Institute of Technology and Business, under project No 8210-008 and 8210-009 and Czech Academy of Sciences, Institute of Theoretical and Applied Mechanics, under project No RVO 68378297.

References

[1] J.M. Chi, R. Huang, C.C. Yang, J.J. Chang, "Effect of aggregate properties on the strength and stiffness of lightweight concrete," *Cem. Concr. Compos.*, vol. 25, pp. 197-205, 2003.
[2] P.S. Lovato, E. Possan, D.C.C.D. Molin, Â.B. Masuero, J.L.D. Ribeiro, "Modeling of mechanical properties and durability of recycled aggregate concretes," *Constr. Build. Mater.*, vol. 26, pp. 437-447, 2012.
[3] B. Cantero, I.F. Sáez del Bosque, A. Matías, M.I. Sánchez de Rojas, C. Medina, "Inclusion of construction and demolition waste as a coarse aggregate and a cement addition in structural concrete design," *Arch. Civ. Mech. Eng.*, vol. 19, pp. 1338-1352, 2019.
[4] L.A.L. Ruiz, X.R. Ramón, S.G. Domingo, "The cilcural economy in the construction and demolition waste sector – A review and an integrative model approach," *J. Cleaner Prod.*, vol. 248, 119238, 2020.
[5] P. Sormunen, T. Kärki, "Recycled construction and demolition waste as a possible source of materials for composites manufacturing," *J. Build. Eng.*, vol. 24, 100742, 2019.
[6] EN 197-1 Cement – Part 1: Composition, specifications and conformity criteria for common cements, CEN, Brussels, 2011.
[7] EN 1015-3, Methods of test for mortar for masonry – Part 3: Determination of consistence of fresh mortar, CEN, Brussels, 2000.
[8] EN 933-1, Test of geometrical properties of aggregate – Part 1: Determination of particle size distribution – Sieving method, CEN, Brussels, 2012.
[9] EN 1015-10, Methods of test for mortar for masonry - Part 10: Determination of dry bulk density of hardened mortar, CEN, Brussels, 1999.
[10] EN 1936, Natural stone test methods – Determination of real density, apparent density and total open porosity, CEN, Brussels, 2007.
[11] EN 1097-3, Test of mechanical and physical properties of aggregate – Part 3: Determination of loose bulk density and voids, CEN, Brussels, 1999.
[12] EN 1015-11, Methods of test for mortar for masonry – Part 11: Determination of flexural and compressive strength of hardened mortar, CEN, Brussels, 2000.
[13] M. Keppert, M. Urbanová, J. Brus, M. Čáchová, J. Fořt, A. Trník, L. Scheinherrová, M. Záleská, R. Černý, "Rational design of cement composites containing pozzolanic additions," *Constr. Build. Mater.*, vol. 148, pp. 411-418, 2017.
[14] EN 206-1, Concrete – Specification, performance, production and conformity, CEN, Brussels, 2018.
[15] Y. Yue, G. Li, X. Xu, Z. Zhao, "Properties and microstructures of plant-figer-reinforced cement-based composites," *Cem. Concr. Res.*, vol. 30, pp. 1983-1986, 2000.
[16] M. Malešev, V. Radonjanin, I. Lukić, V. Bulatović, "The effect of aggregate, type and quality of cement on modulus of elasticity of lightweight aggregate concrete," *Arab. J. Sci. Eng.*, vol. 39, pp. 705-711, 2014.

IOP Conf. Series: Materials Science and Engineering 1066 (2021) 012009 doi:10.1088/1757-899X/1066/1/012009

Influence of Amount of Filler on Thermal Resistance of Bitumen Sheets

Jan Plachý [1], Jana Vysoká [1], Radek Vejmelka [1]

[1] Institute of Technology and Business in České Budějovice, Okružní 517/10, 370 01 České Budějovice, Czech Republic

plachy@mail.vstecb.cz

Abstract. The paper deals with the correlation between the amount of fillers and the thermal resistance of bitumen sheets under EN 1110:2011. Based on the analysis of samples of bitumen sheets used for the insulation of concrete bridges, the standard EN 14695:2010 indicates what the usual amount of fillers is and what the value of the resistance to flow at elevated temperature is. Bitumen sheets, which are used in the Czech and Slovak Republic, were used as samples. The samples represented bitumen sheets modified with polymers of elastomeric and plastmeric nature. Bitumen sheets were of two types with and without protective covering. It was found that fly ash and micro-milled limestone were used as fillers in the bituminous substance of the sheets. The dependence between the amount of fillers and the resistance to flow at elevated temperature was only confirmed in some of the bitumen sheets samples.

1. Introduction

Bitumen sheets are the important elements of the building industry due to their waterproofing properties. The quantity and composition of bitumen coating determine the waterproofing function of bitumen sheets. The important component of the bituminous substance is given by the filler. The well-known is the fact that an increased quantity of the filler decreases the cost of bitumen and then this fact influences the cost of bitumen sheets. Quantity of the bitumen filler is mostly determined by the technological potentialities of the production line. However, the quantity of the filler could negatively influence some parameters of the bitumen mass and hence the bitumen sheets. One of the basic properties influencing a functionality of bitumen sheets is the flow resistance at elevated temperatures [1]. The increased amounts of fillers also decrease the on-site application.

This paper is focusing on the analysis of the amount of filler in the bituminous mass of waterproofing sheets used for the insulation of concrete bridges [2]. The contribution describes usual amount of fillers used in the bituminous mass of waterproofing sheets from different producers. And it describes values of resistance to flow at elevated temperature according to modification of bituminous mass.

During the previous research, the authors of this paper came to the conclusion that the relation between the quantity of the filler and the flow of resistance at elevated temperatures is determined both by the type of bituminous substance and protective covering [3]. It has been statistically proved that the main factor is not the amount of the filler but the type of bitumen mass.

The main aim of this article is to confirm or confute the hypothesis: Is there any dependency between the flow resistance at elevated temperatures and amount of fillers? The hypothesis was tested on an extended set of samples. Thus, new results can confirm or confute previous research conclusions. Results were summarized based on the statistical evaluation of data collected from laboratory measurements.

The information available to us shows that the topic of fillers in bitumen mass of waterproofing sheets was dealt only by [4]. However, the study concerned the manifestation of the aging of the bitumen mass. The amount of fillers in the bitumen mass is reflected in a reduction in the strength of the joints. Decreased strength in joints leads to failures, which are especially dangerous in single-layer flat roof systems [6]. The issue of recycled plastic fillers is studied by [7].

2. Material and methodology

2.1. Used material

The test selection included thirty-four samples used for monolayer applications for the insulation of concrete decks in the Czech and Slovak Republic.

Nineteen of the samples represented samples with coarse grained gritting, nine had fine grained gritting and six samples were without gritting. There were fourteen samples with the bitumen mass modified by elastomers and twenty samples with the mass modified by plastomers. All samples contained a reinforcement layer made of polyester impregnated by the bitumen mass. Basic bitumen sheets characteristics are summarized in Table 1. Numbering of samples was created during the diagnosis of bridge structures in the Czech and Slovak republic, some samples were not suitable for further testing, therefore their numbers are missing in the number series.

Table 1. Selected characteristics of tested samples. Where HP – coarse gritting, JP – fine-grained gritting, N – the finish of the geotextiles weighing about 200 [g/m^2], PES – polyester fleece , EP – mass modified by elastomers, P – mass modified by plastomers, A – under the surface, in the upper 1st twelfth of the thickness of the sheet, B – in the upper 2nd -3rd twelfth of thickness of the sheet, C – in the upper 4th-5th twelfth, D – in the upper 6th twelfth of thickness of the sheet [8] and authors.

Sample number	Protective covering	Type of bitumen mass modification	Reinforcement - area weight [g/m^2]	Sheet thickness [mm]	Reinforcement Position	State
2	HP	P	230	5.20	C	CZ/SK
3	N	P	200	5.24	A	CZ/SK
4	HP	E	230	5.30	A	SK
6	N	P	200	4.83	A	SK
7	JP	E	250	5.10	C	SK
8	HP	P	220	5.50	C	SK
9	JP	P	230	5.16	B	CZ/SK
10	HP	P	230	5.16	B	CZ/SK
14	JP	P	230	6.25	D	CZ
15	HP	E	230	5.46	C	CZ
16	JP	E	230	4.80	D	CZ
17	HP	P	230	5.67	C	CZ
18	HP	E	250	5.05	B	SK
22	HP	P	250	5.22	A	SK
23	HP	P	280	4.80	C	CZ
25	HP	E	280	5.18	B	SK
26	N	P	250	4.75	A	SK
28	HP	P	250	5.30	A	SK

29	N	P	250	5.04	B	CZ
30	HP	P	250	4.81	B	CZ/SK
31	HP	P	280	5.23	B	CZ/SK
32	HP	E	250	5.40	C	SK
33	HP	E	250	5.34	B	SK
34	HP	E	230	4.81	C	CZ
35	JP	E	250	4.99	B	CZ
39	N	P	280	5.12	D	CZ
40	N	P	280	5.00	A	CZ
41	N	E	230	5.41	C	CZ
42	HP	P/E	230	5.41	C	CZ
43	HP	P	250	4.90	A	CZ
44	HP	P/E	230	4.97	C	CZ
45	HP	E	230	4.50	C	CZ
46	HP	P	250	4.92	C	CZ
47	JP	P	230	5.19	A	CZ/SK

2.2. The thermal resistance of bitumen sheets
The test of thermal resistance was executed according to [1]. The type and the quantity of the filler were determined on the basis of the previous measurement [8]. Amount of fillers was determined in [9]. The tested bitumen sheets were separated into two groups in terms of the character of the filler – limestone and ash. In case of other type of filler, the sample would be categorized to new group.

2.3. Methodology of statistical evaluation of influence of filler amount on thermal resistance
Evaluation for three categories was implemented: A – up to 20 % inclusive, B – more than 20 % and maximally 30 %, C – more than 30 % of the quantity of the filler because the quantity of the filler is generally a numerical variable. Kruskal-Wallis test was carried out as the nonparametric statistical test due to the nature of the data and the parametric test (ANOVA) were used for the result comparison of results. However, an importance of nonparametric tests is more significant due to the nature of data. The evaluation of all tests was reached at the significance level with 95% confidence, i.e. with the coefficient $\alpha = 0.05$. The limits of 20 and 30% of fillers were determined on the basis of the usual amount of fillers in bitumen, which results from previous research [8].

3. Test results
3.1. Determination of the flow resistance at elevated temperatures according to EN 1110:2011
Bitumen sheets with mass modified with elastomers reached values from 105 °C to 130 °C, the mean value is given by the value 117.5 °C and the standard deviation is 8.4. Results of testing of sheets with the mass modified by plastomers reached values in the limits from 110 °C to 160 °C, the mean value is calculated as the value 145 °C and the standard deviation is 16.81. Table 2 collects complete results.

3.2. Determination of amount and type of fillers in bituminous mass
The quantity of fillers in the bitumen mass oscillated from 9.7% to 50,9% with average value 27.7 % and standard deviation is 8.3%. Fly ash was used in 16 bitumen sheets with average value of 26.3% and standard deviation of 4.6%. Quantity of fly ash in the bitumen mass oscillated from 19.9% to 38.3% . Limestone was used in 18 bitumen sheets with average value of 23.9% and standard deviation of 10.8 %. Table 2 collects complete results

Table 2. Results of determination of the flow resistance at elevated temperatures according to [1], amount and type of fillers for appropriate samples [3], [8], authors. Where V – Limestone, P – Ash, Source: authors.

Sample number	Flow resistance at elevated temperatures [°C]	Filler amount	Type of filler
2	155	14.1	V
3	155	18.7	V
4	120	24.1	V
6	115	24.6	P
7	125	21.3	P
8	120	25.4	P
9	160	19.7	V
10	160	20.0	V
14	125	21.7	P
15	100	26.0	P
16	110	24.4	P
17	125	26.7	P
18	130	35.0	V
22	160	13.0	V
23	160	14.6	V
25	125	30.9	V
26	140	19.9	P
28	110	26.2	V
29	155	22.3	V
30	150	23.2	P
31	155	38.3	P
32	125	50.9	V
33	115	31.6	V
34	115	40.0	V
35	110	35.3	V
39	155	22.3	V
40	155	9.7	V
41	120	28.9	P
42	125	26.5	P
43	155	10.1	V
44	120	27.3	P
45	105	30.8	P
46	130	22.7	V
47	160	24.7	P

3.3. Statistical evaluation of influence of quantity of filler on resistance to flow at elevated temperatures

Test results give the value of the test statistics and p-value using the methods ANOVA and Kruskal-Wallis test with values 11.82 and 0.000153, 13.588 and 0.001121. Other results are shown in Figure 1. Where the graph of Tukey's multiple comparisons of mean values of resistance to flow at elevated temperatures is presented.

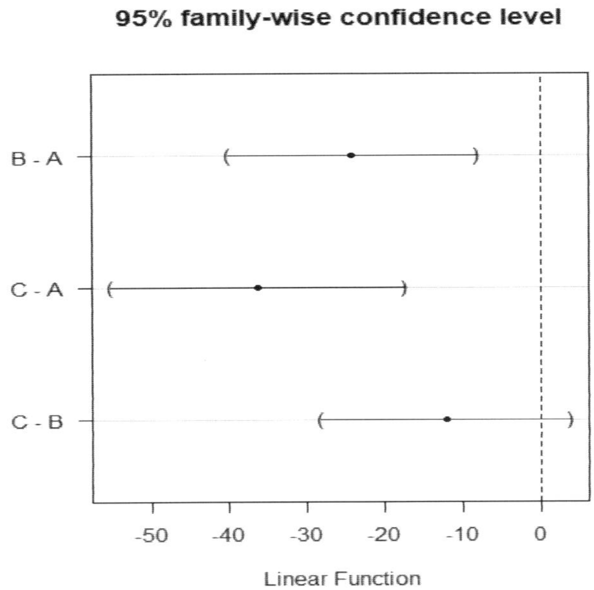

Figure 1. The graph of Tukey's multiple comparisons of mean values of resistance to flow at elevated temperatures. Source: authors.

4. Results and discussions

4.1. Resistance to flow at elevated temperatures, determination of amount of filler and type of filler in bituminous substance of waterproofing sheets

Bitumen sheets are divided to groups according to modification of bituminous substance and type of filler used. Results are arranged in synoptic chart, see figure 2.

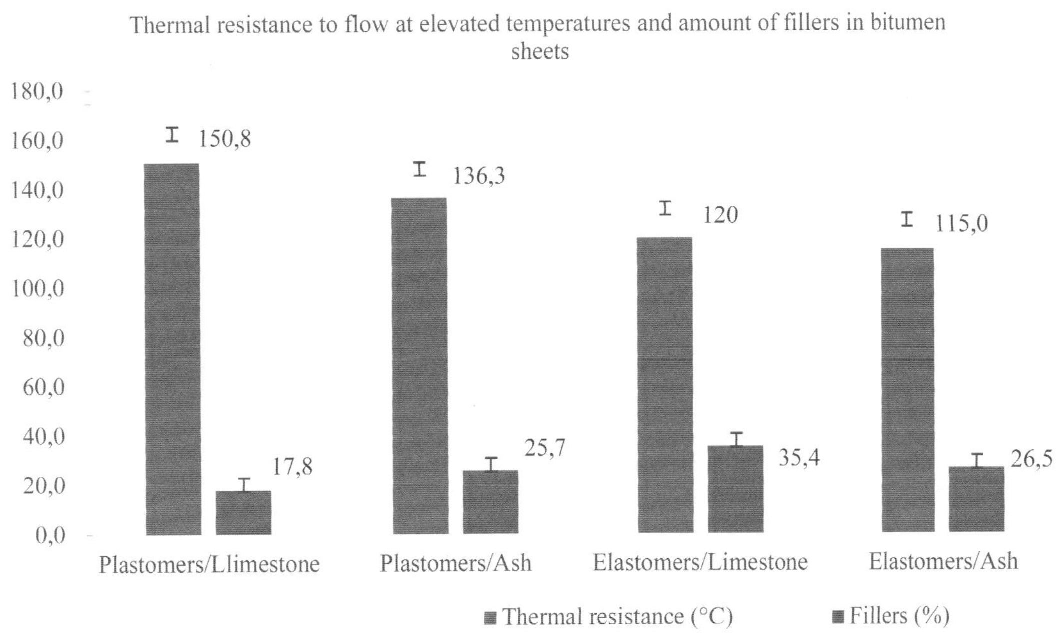

Figure 2. Thermal resistance to flow at elevated temperatures and amount of fillers in bitumen sheets. Source: authors.

The above chart shows that:

- Bitumen sheets modified by plastomers have higher resistance to flow at elevated temperatures. This result confirms their use. Bitumen sheets modified with plastomers are used for applications with more thermally stressed.

- In terms of bituminous mass, there is a higher amount of fillers in bitumen sheets modified with elastomers by about 10%. This result is caused by the workability of the bitumenous mass during production. Bitumen sheets modified by elastomers contain smaller amounts of polymer than bitumen sheets modified by plastomers. The bitumen mixture thus has a lower viscosity and it is thus possible to add a higher amount of fillers.

- A higher amount of fillers increases the heat resistance only for strips modified with elastomers. The reason is again a lower amount of polymer in the bitumenous mass.

- The weight dosage of fly ash into the bituminous mixture is comparable regardless of the type of asphalt mass. The dosage of limestone differs. This is caused again probably due to the workability of the bitumenous mixture. It can be explained by the fact that the bulk density of fly ash (800 kg/m3) is lower than that of limestone (1400 kg/m3) [10]. In terms of volume in the bitumenous mass, limestone occupies a smaller volume than fly ash.

- Bitumen sheets where limestone was used as filler have higher heat resistance. Limestone particles are sharp-edged that is a more suitable than spherical fly ash in terms of resistance to flow.

4.2. Results of statistical evaluation

The results of statistical processing show that the amount of fillers has a statistically significant effect on the resistance to flow only for category A (amount of fillers up to 20%). The p-value of all tests is less than the significance level of 0.05. And the chart shows that the interval estimation of the difference between the mean values of resistance to flow according to the categories of filler amount between category A and other categories is significantly non-zero. There is no statistically significant difference between categories B and C. From the result it can be deduced that the addition of fillers changes the properties of the asphalt mixture, there is an increase in heat resistance. The filler stabilizes the bitumen mixtures at elevated temperatures. However, as the amount of fillers increases, this effect decreases.

5. Conclusions

The amount of fillers in plastomer modified bitumen sheets is 21.7% and in elastomer modified sheets is 30.9%. Results show that the filler has an effect on the resistance to flow at elevated temperatures of bitumen sheets modified with elastomers.

From a statistical point of view, bitumen sheets with limestone as filler have a higher resistance to flow at elevated temperatures.

The hypothesis assuming that the type and amount of filler have an effect on the resistance to flow at elevated temperature was confirmed only for bitumen sheets with the amount of fillers up to 20%. Thus, a small amount of filler will affect the flow resistance. With higher amount of fillers, the effect is lost.

References

[1] EN 1110:2011. Flexible sheets for waterproofing. Bitumen sheets for roof waterproofing. Determination of flow resistance at elevated temperature. Prague: Czech Standards

Institution. 2011-05-01. Classification mark 727634
[2] EN 14695:2010. Flexible sheets for waterproofing – Reinforced bitumen sheets for waterproofing of concrete bridge decks and other trafficked areas of concrete – Definitions and characteristics, Prague: Czech Standards Institution. 2010-05-01. Classification mark 727605. 2010
[3] J. Plachy, J. Vysoká, R. Vejmelka, V. Petránek, T. Petriček, L. Kalousek and P. Bednářová, "The Statistical Evaluation of Thermal Resistance and Price of Bitumen Waterproofing Sheets," In International Conference on Numerical Analysis and Applied Mathematics 2014, *ICNAAM 2014. 1. edition. Melville (New York): American Institute of Physics Publishing LLC*, not paged, 4 pages, 2015.
[4] S. Xu, W. Da, W. Li, and J.Y. Yu, "Effect of fillers on aging properties of LDHs/SBS modified asphalt coating," *Wuhan Ligong Daxue Xuebao/Journal of Wuhan University of Technology,* vol. 35, pp. 41-44, 2013.
[5] M. Garrido, D. António, J. G. Lopes, and J. R. Corrieia, "Performance of different joining techniques used in the repair of bituminous waterproofing membranes, " *Construction and Building Materials*, vol. 158, pp. 346-358, 2018.
[6] T. Petříček, P. Kacálek, and T. Hlavačka, "Characteristics of the mechanically fastened joints of one-layer bitumen sheets, " MATEC Web of Conferences, vol. 93, 22 December 2016, Article number 03007, *8th International Scientific Conference on Building Defects, Building Defects 2016*, Institute of Technology and Business (ITB) Ceske Budejovice, not paged, 6 pages, 2016.
[7] R. Smolka, and J. Sobotka, "Application of recycled plastic in flat roofs," MATEC Web of Conferences, vol. 146, 22 January 2018, Article number 02009, *9th International Conference Building Defects, Building Defects 2017*, Institute of Technology and Business (ITB) Ceske Budejovice, not paged, 6 pages, 2017.
[8] J. Plachý, J. Vysoká, and R. Vejmelka, "Influence of the Quantity of Fillers on Crucial Thermal Technical Parameters of Bitumen Waterproofing Sheets," *IOP Conference Series: Materials Science and Engineering,* vol. 471, Issue 9 (World Multidisciplinary Civil Engineering-Architecture-Urban Planning Symposium 2018). 1. edition United Kingdom: Institute of Physics Publishing, not paged, 7 pages, 2019.
[9] ČSN 50 3602:1967. Testing of roofing and waterproofing materials in roles. 1st ed. Prague: Czech Standards Institution, 1967-02-22. Classification mark 503602.
[10] Z. Plecháč and coll. ABC of bitumen waterproofing sheets. 2019. Association of bitumen sheets manufacturers of the Czech Republic, i.a. Prague. Accesible from: https://svaz-ap.cz/content/ke-stazeni/publikace

Problems of Regulating the Safe Evacuation of People at the Subway Stations

A Parfyonenko [1], E Kostyuchenko [1]

[1] National Research Moscow State University of Civil Engineering, Yaroslavskoye Shosse, 26, Moscow, 129337, Russian Federation

ikbs@mgsu.ru

Abstract. This article describes the main provisions for assessing the safety of people during the evacuation from the subway stations. A huge number of passengers waiting for the train, leads to a seal flow. The high-density value reduces the width of the doorway in the light. People tend to enter the car without waiting for the passengers to leave it, being afraid not to have time to land in the allotted time for this or not to get into the car as a result of its overfilling. Higher intensity values can be observed in the case of a small number of waiting passengers. The density of people on the platform when waiting for the train determines, in addition, the comfort of the waiting conditions of passengers and their safety. The presented results of the study on the safety of subways will allow more accurately and differentially determine the movement of human flows in the different sections of the track at the metro stations.

1. Introduction

According to the Town Planning Code of the Russian Federation (Part 1, article 48. 1) [1] subways are considered to be particularly dangerous, technically complex facilities. Therefore, the safety of people in pedestrian subway structures (stations and interchange nodes) should be determined by calculation based on an assessment of individual fire risk, according to the Federal Law of the Russian Federation of 22.07.2008 No.123-FZ "The technical regulation about requirements of fire safety"[2]. The requirements of SP 120.13330.2012 "Subways" [3] also comply with this concept, that sufficiency of design decisions to ensure the safe evacuation of people at the station must be estimated by calculation, and when calculating the evacuation time, all protected escape routes are taken into account. Further, paragraph 5.4.1.3 indicates that the length of the no-break sections at the ends of the landing part of the platforms of the deep-laying station should be taken no more than 1/3 of the length of the landing platform and determined from the conditions that the passengers should be released from this section during a period not exceeding the minimum interval between by trains and within the estimated time of evacuation of passengers from the station. However, neither SP 120.13330.2012 "Subways", nor methodology for the determining the estimated values of fire risk in buildings, structures and structures of various classes of functional fire hazard [4] give the calculated values of the parameters necessary to determine the probability of the evacuation.

2. Methodology

2.1. Evacuation routes and exits

The paragraph 5.16.6.3 of SP 120.13330.2012 "Subways" indicates that the following routes should be provided for evacuation from the platform halls of the station:

1. Along escalators and (or) stairs of the 2nd type, corridors, through the cash rooms of the lobby, underground passages - until they exit;
2. Through interchange facilities - to the station of another line and further as in point "1", thereby expanding the concept of escape routes and exits.

2.2. Estimated fire hazardous situation

The SP 120.13330.2012 "Subways" does not establish the design situation for determining of sufficiency of design decisions to ensure the safe evacuation of people at the station. A possible place of occurrence of a fire source may be a household or industrial premises, one of the rooms for the movement of passengers, a wagon moving in a train tunnel. The paragraphs of SP 120.13330.2012 "Subways" establish:

- The paragraph 5.16.6.1: "... In the event of a fire in one of the wagons of a train moving in a tunnel, it must continue to move to the nearest station to evacuate people and extinguish the fire";
- The paragraph 5.16.6.4: "When equipping rolling stock elements (carriage equipment, hardware compartment, driver's cab) with an automatic fire extinguishing installation, fire-fighting is carried out while the train is moving. Moreover, the evacuation of passengers is carried out after the arrival of the train at the station".
- As a calculated fire hazard situation, a fire should be taken in one of the train cars located at the station.

2.3. Estimated evacuation scheme

The paragraphs of SP 120.13330.2012 "Subways" require:

- The paragraph 5.16.6.17: "... The estimated number of people (including MGN) in the station facilities must be determined based on the maximum prospective passenger flows of the designed station and the conditions that trains follow the non-emergency path without stopping at the station with fire (to the passage)";
- The paragraph 5.16.6.18: "To ensure the safe evacuation of people in case of fire, when the fire alarm system is activated, turnstiles must be unlocked with free movement of people in both directions";
- The paragraph 5.16.6.2: "From the platform halls of the station, at least two dispersed evacuation exits should be provided, ensuring safe evacuation of people in case of fire".

The ability to block one of the evacuations exits of the platform hall of the station determines the unidirectional movement of human flows during evacuation. Sources of human flows are exits from burning cars and station platforms with passengers on them waiting for the arrival of the next train.

The estimated number of evacuated is determined by the maximum congestion during peak hours and the congestion of both station platforms by waiting passengers. The design evacuation scheme has a general view, presented on Figure 1.

IOP Conf. Series: Materials Science and Engineering 1066 (2021) 012010 doi:10.1088/1757-899X/1066/1/012010

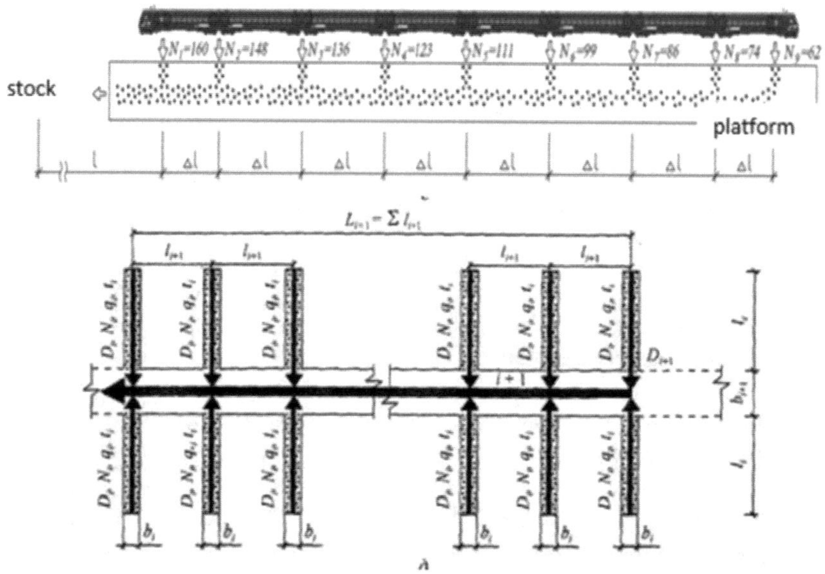

Figure 1. The estimated evacuation scheme.

2.4. Features of the movement of human flows in areas in the estimated evacuation scheme

The features of the movement of human flows in areas in the estimated evacuation scheme include:

- exit of people from cars;
- flow formation on the platform;
- expanding flow on a section of a common path;
- the movement of human flows in areas in front of escalators associated with the functioning parameters of escalators;
- the movement of human flows through automatic checkpoints.

The description of these features required specially organized field observations [5-19], the full complex of which was first performed by the video method [6, 7, 20] on Figure 2.

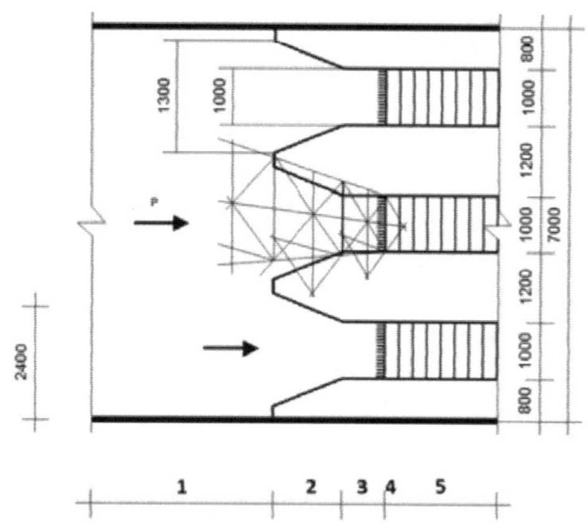

Figure 2. The example of conducting field observations using the method of video observations with the construction of a computational grid: 1 - the first section (horizontal path); 2- second section (horizontal path between the guiding devices); 3- third section (horizontal path between balustrades); 4 - the fourth section (comb); 5 - the fifth section (escalator tape).

2.5. People leaving from the metro coaches

The need to study human flows on the platform of the metro station is associated with the performance of two functions:

- the formation of the people leaving from the metro coaches,
- accumulation of passengers while waiting for the train.

The field observations of the movement of people on the platform were carried out at "Turgenevskaya" Moscow metro station under normal operating conditions of the metro and during "peak" hours.

The values of the intensity of the people leaving from the metro coaches are presented in Table 1.

Table 1. The values of the intensity of the people leaving from the metro coaches.

Observation period	Number of observations	Intensity, q people / m min	Dispersion, S^2 people^2/ m^2 min^2
Summer	125	49.2	118.02
Winter	119	51.13	161.93

The above results indicate that there is no influence of the time of the year on the intensity of people leaving from the metro coaches.

2.6. People flow formation on the platform

A large number of passengers waiting for the train, leads to a seal flow. The density of people in front of the car door when the train approaches reaches 5 people/m^2. A high-density value reduces the width of the doorway in the light. People tend to enter the car without waiting for the passengers to leave it, being afraid not to have time to land in the allotted time for this or not to get into the car as a result of its overfilling. Higher intensity values can be observed in the case of a small number of waiting passengers.

The density of people on the platform when waiting for the train determines, in addition, the comfort of the waiting conditions of passengers and their safety. The density value should be taken into account when determining the area of the "waiting zone" of the station platform. After people exit the cars and from the platform to the distribution room of the metro station, the width of the human flow increases [6,9-13, 18], the so-called "expansion" of the flow (Figure 3).

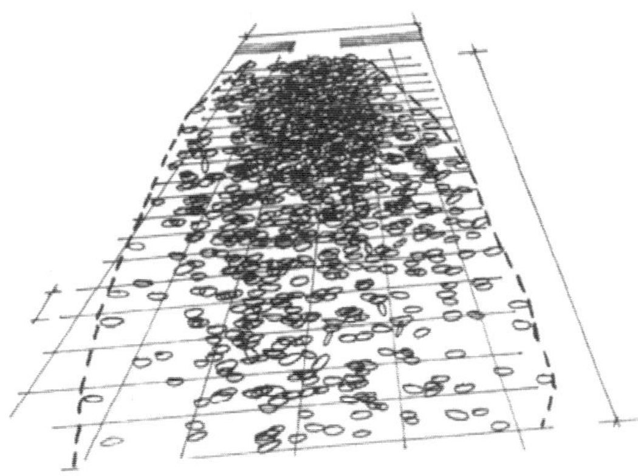

Figure 3. The expanding flow scheme.

The reason for the formation of "expanding" human flows is the desire of participants to go with a lower density, which allows to increase speed and increase comfort of movement. Each person seeks to reduce the time spent on movement, and this can be achieved in two ways: the first is an increase in the speed of movement, which occurs due to a decrease in the flux density; the second is the reduction of the path length. The latter is the reason that the human flow stops expanding due to the deviation of the trajectories of movement from the shortest direction. The stabilization of the width of the flow occurs when it reaches a certain value of its density, at which a certain equilibrium is established between the density, speed and path length.

The obtained probability distribution of the density values of "expanding" flows is shown on Figure 4.

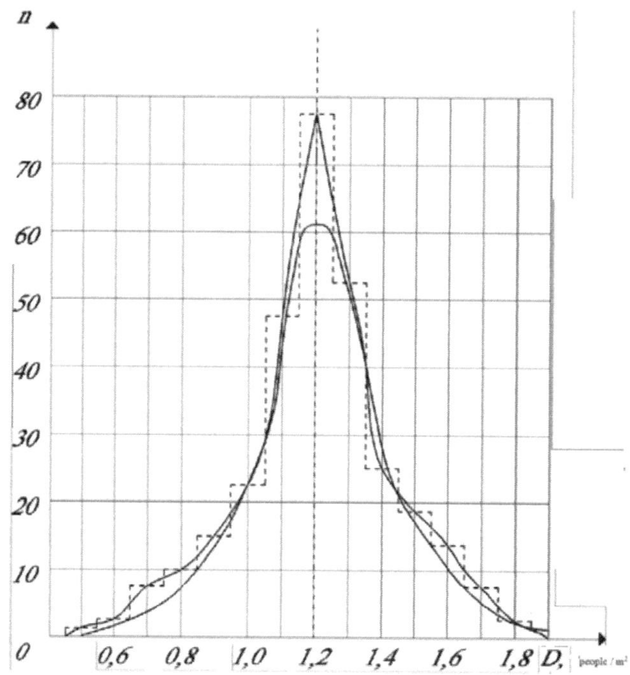

Figure 4. The density distribution approximation of the expanding flows by the normal law of distribution:

- - - - - histogram of statistical distribution;
———— statistical distribution latitude;
▬▬▬▬ theoretical distribution curve.

The regularity of the relationship between the speed and density of the human flow [8,15] is described as an elementary random function:

$$V_D = V_0(1 - a\ln\frac{D}{D_0}) \quad (1)$$

where V_D is the value of the random variable velocity at the flux density D, which depends on the type of path and on the psychological tension of the movement process; V_0 is a random value of the flow velocity at a flux density lower than D_0; D_0 is the threshold value of the density, upon reaching which it becomes a factor that affects the speed of people in the stream.

IOP Conf. Series: Materials Science and Engineering 1066 (2021) 012010 doi:10.1088/1757-899X/1066/1/012010

As can be seen, the values of the coefficient for both types of motion are very close. Since in both cases the movement occurs along a horizontal path, it is legitimate to accept a single coefficient value equal to the average value. It amounted to a = 0.3997 ≈ 0.4000.

2.7. The movement of people flows in areas in front of escalators

Based on the results of field observations, conclusions were drawn that the speed of the escalator bed affects the parameters of the movement of the human stream. An increase in the velocity V and the intensity of movement q at the same flux density with a decrease in the velocity of the escalator to V_e= 0.74 m/s and a decrease in V with a further decrease in V_e were recorded.

Attention is drawn to the identity of changes in the parameters of the movement of the human flow when the velocity of the escalator bed deviates both to a smaller and a larger side from the value of V_e= 0.74 m/s. So, for example, the parameters of the movement of people at V_e= 0.56 m/s are close in their values to the parameters of movement at V_e = 0.84 m/s. This pattern is explained by an increase in the concentration of people in front of the escalator with an increase or decrease in the speed of the canvas. At small and large values of V_e, the cluster spread over almost the entire length of the visible section of the path.

How quickly the transition through the comb will occur, whether a movement delay occurs and how long it is, depends on how much the approach speed V_a differs from the speed of the escalator V_e:

$$q(V_e) = -a(V_e - V_a)^2 + q_{max}. \quad (2)$$

3. Results and discussions

The results of the field observations showed that the carrying capacity of the escalator Q_e increases with an increase in its speed to 0.7 m / s, and then it decreases. The reason is that at an escalator speed of V_e = 0.7 m/s = 42 m/min, the difference between V_a and V_e is minimal, this ensures the greatest value of the carrying capacity of the escalator.

The approximation of the dependence of the carrying capacity of the escalator on the speed of movement of his canvas is determined by the formula:

$$Q(V_e) = 60\{19,6 + [-0,028(V_e - 42,37)^2]\}/f, \text{ man/hour} \quad (3)$$

Field observations [5, 6, 9, 14, 16, 17] of the movement of people flows through automatic checkpoints during peak hours under normal operating conditions with flux densities in front of them from 1 to 5 people / m^2 show that their transit time (t_{pr}) depends from the tension of the movement process. At D = 2 people / m^2, the value of t_{pr} under normal conditions, due to an increase in traffic intensity, approaches the value of t_{pr} during peak hours (Figure 5).

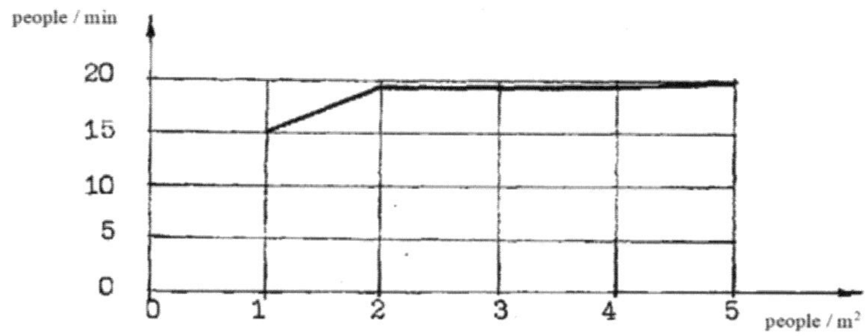

Figure 5. The dependence of the transmission capacity of the automatic transmission on the density of people flow in front of them.

4. Conclusions

The stated research results provide the necessary prerequisites for the development of the code of rules "Fire safety of subways" and the development of the methodology for determining the calculated values of fire risk in buildings, structures and structures of various classes of functional fire hazard [4], which will make it possible in the future to more accurately and differentially determine the parameter P_e, taking into account the movement of people flows through different sections of the track at metro stations and to develop reasonable fire-fighting measures.

References

[1] The Town-Planning Code of the Russian Federation dated 29.12.2004 No.190-FZ (in Russian), 2004.
[2] Federal Law of the Russian Federation of 22.07.2008 No.123-FZ "Technical regulation on fire safety requirements" (in Russian), 2008.
[3] SP 120.13330.2012 "Subways". With amendment No. 1 and amendment No. 2 (in Russian).
[4] The methodology for determining the calculated values of fire risk in buildings, structures and structures of various classes of functional fire hazard. Appendix to the Order of the Ministry of Emergencies of Russia dated June 30, 2009 No. 382, subject to the changes approved by Order No. 632 of the Ministry of Emergencies of the Russian Federation of December 2, 2015 (in Russian), 2009.
[5] V. Kholshchevnik, "The influence of patterns of urban population movement on the throughput of access points of multifunctional high-rise buildings-complexes," *Fire and explosion safety,* vol. 17, no 1, pp. 34-43 (in Russian), 2008.
[6] I. Isaevich, *Development of the foundations of a multivariate analysis of planning decisions of stations and interchange nodes of the subway based on modeling the laws of movement of human flows*, Dis. Ph.D. Moscow (in Russian), 1990.
[7] V. Kholshchevnikov, D. Samoshin, and I. Isaevich, F*ield observations of human flows: textbook.* allowance. - M.: Academy of State Fire Service of the Ministry of Emergencies of Russia (in Russian), 2009.
[8] V. Kholshchevnikov, *The patterns of communication between the parameters of human flows.* Diploma No. 24 - S / V.V. Kholshchevnikov. - M.: Russian Academy of Natural Sciences, International Academy of Authors of Scientific Discoveries and Inventions (in Russian), 2005.
[9] V. Kholshchevnikov, D. Samoshin, and I. Isaevich, *Recommendations for the design of pavilions for automated access control (ASKD) at railway stations, taking into account the movement of human flows*, Research report - M.: AGPS EMERCOM of Russia (in Russian), 2005.
[10] A. Milinsky, *Study of the evacuation of buildings for public use*, Dis. Ph.D. 05.23.10 Moscow 178 p (in Russian), 1951.
[11] O. Aydn, "A Qualitative Approach to Children of Developing Countries from Human Behavior

Point of View," *Proceedings of the 2nd International Symposium on Human Behaviour in Fire*, pp. 531 – 538, 2011.

[12] D. Bruck, "Community based research on the effectiveness of the home smoke alarm in waking up children," *Proceedings of the 4th International Symposium on Human Behaviour in Fire*, pp. 335 – 344, 2009.

[13] A. Larusdottir, And A. Dererichs, "Evacuation dynamics of children – walking speed, flow through doors in day care centers," *Proceedings of the 5th International Conference on Pedestrian and Evacuation Dynamics*, pp. 67 – 79, 2010.

[14] Y. Alekseev, *Formation and movement of human flows in the aisles of spectacular structures*, Dis. Ph.D. 05.23.10 Moscow (in Russian), 1978.

[15] V. Predtechenskiy, *Human Streams in Buildings*, Moscow: Stroyizdat (in Russian), 1983.

[16] V. Kalintsev, *Designing cinema halls taking into account the movement of human flows*, Dis. Ph.D. 05.23.10 Moscow (in Russian), 1966.

[17] R. Duvidzon, *Design of sports facilities taking into account the movement of human flows*, Dis. Ph.D. 05.23.10 Moscow (in Russian), 1968.

[18] R. Grigoryants, *The study of long-existing human flows*, Dis. Ph.D. 05.23.10 Moscow (in Russian), 1973.

[19] V. Gvozdyakov V, *Patterns of movement of human flows in transport and communication facilities*, Dis. Ph.D. 05.23.10 Moscow (in Russian), 1978.

[20] V. Kopylov, *Research of parameters of the movement of people during forced evacuation*, Dis. Ph.D. 05.23.10 Moscow (in Russian), 1974.

Ductwork Pressure Loss Determination Utilizing Building Information Model

Michal Nováček [1], Josef Remeš [1]

[1] Institute of Computer Aided Engineering and Computer Science, Faculty of Civil Engineering, Brno University of Technology, Veveří 95, 602 00 Brno, Czech Republic

xpnovac03@vutbr.cz

Abstract. Building design created as an information model contains not only information formated during building geometry creation but also additional information filled in by a designer. The data can be used further as well as applied in building geometry analyses. One of these are analyses of pressure loss in air-handling units. Such analyses take the advantage of both created geometric parameters, and basic physical variable. These analyses can be carried out directly in BIM software without external calculations needed. The main aim of this article is to describe utilization of information in building model leading to obtain pressure loss in air-handling units.

1. Introduction

Building created as a building information model provides plenty of opportunities. Information contained in model are as comprehensive as the designer is. The model contains fundamental geometric information which become the standard in civil engineering. On the other hand, building information model can contain complementary information. These complementary information are especially physic origin and can be utilized for building analysing which is important for sustainable building design. Project design phase is a building life cycle phase when the most important decisions are made, and changes occur. The decisions are important for further realization and building using phase especially because of expenditure. Over time decisions converge to an optimal solution.

In some cases, the building can be analysed directly into the modelling software with no external applications or other methods needed. Commercial software already provides analysing tools utilizing physical models. However, most of the analyses are not possible to be utilized directly in modelling software and one of them is ventilation ductwork pressure loss analysis. The analysing tools are influenced by building national standards requirements. Although physical equations are general, its interpretation and utilizing depends on national habits and standards. That is the reason why inconsistency is in the analysing tools between countries. The solution can be to develop national standard analysing tools with the use of BIM data in building design phase. [1]

2. Ventilation ductwork pressure loss

One of the requirements for proper ventilation ductwork design is a fan power sufficient to supply air into every room. [2] Ductwork pressure loss is needed to be calculated for designing the fan. Pressure losses are caused by air friction against the duct walls [3]. The pressure loss on straight duct sections is very small. On the other hand, in places where flow direction is changed (elbows, reductions, other ventilation fittings), the pressure loss is considerable. For pressure losses in duct fittings (excluding straight pieces of duct) we use the term local pressure loss.

Local pressure losses are expressed as a pressure drop by the difference of static pressures at the beginning of the fitting and at its end. This is based on Bernoulli's equation

$$\rho \frac{v_1^2}{2} + p_1 + \rho g h = \rho \frac{v_2^2}{2} + p_2 + \rho g h + p_z, \tag{1}$$

where ϱ is the air density, v is the mean velocity in a given cross-section, p is the static pressure, h is the height difference between the measured cross-section and the reference plane and p_z is the pressure loss caused by friction.

It is assumed that the system is airtight, and the mass flow is therefore constant. If the duct cross-sectional area is constant, the mean flow velocities in different parts of the duct are equal. In the case of air ducts, the effect of gravity on the flow is often neglected. With this simplification, the pressure drop of the duct fitting can then be expressed as equation

$$p_z = p_2 - p_1 \tag{2}$$

This pressure drop can also be expressed in terms of mean velocity as

$$\Delta p = \xi \rho \frac{v^2}{2}, \tag{3}$$

where ϱ is the air density, v is the flow mean velocity and ξ is the pressure loss coefficient.

The pressure loss coefficient is based on the duct fitting type and depends especially on the nature of the flow (Reynolds number) and the fitting geometry. The pressure loss coefficient values can be found in tables [3-5] or (less often) in the form of mathematical equations [3, 6-7].

For pressure loss calculation on straight ducts sections is used equation

$$\Delta p = \lambda \cdot \frac{l}{D} \cdot \frac{\rho v^2}{2}, \tag{4}$$

where λ is the friction factor, l is the length of the straight duct section, D is the duct hydraulic diameter, ϱ is the air density, v is the flow mean velocity.

The Colebrook-White equation is often used for the friction factor calculation

$$\frac{1}{\sqrt{\lambda}} = -2 \log\left(\frac{2{,}51}{Re\sqrt{\lambda}} + \frac{\varepsilon}{3{,}7\, D}\right), \tag{5}$$

where ε is the absolute roughness of the duct material, D is the duct inner diameter and Re is the Reynolds number. In addition to the Colebrook-White equation, its approximation equations are also often used. These equations express the friction factor explicitly [6].

To select the sufficient fan power, it is necessary to calculate the total ductwork pressure loss. That means to sum the pressure losses of all fitting types and straight duct sections on the ductwork. The designer is forced to consider various geometric parameters and types of individual fittings to select the appropriate pressure loss coefficient. However, this process can be automated utilized a ductwork model in building information model.

3. Building information model

Plenty of software that provides modelling buildings as BIM, also allows to model building services. The mere geometry of a given duct section expressed by parametric values can serve as an input variable in the calculation of the pressure loss.

Some software, such as Autodesk Revit, allow to analyse ventilation ducts directly in the application's user interface. Using mentioned software, it is possible to analyse the ventilation ducts for pressure loss by means of several methods. For straight duct sections it is possible to choose from a couple of approximation equations (specifically the Haaland equation and the Altshul-Tsaal equation) and the Colebrook-White equation (5) to obtain the friction factor. The internal calculation also considers the nature of the flow according to the value of Reynolds number (software distinguishes between laminar and turbulent flow). The total straight duct sections pressure loss is determined using the modelled geometry and specified amount of supplied air.

The duct fitting pressure loss is calculated according to equation (3) utilizing the mass flow and the pressure loss coefficient. The pressure loss coefficient is obtained in several possible ways: by the input of user-entered coefficient value for each entity (each duct fitting) or by the user duct fitting pressure loss input directly to each entity and or by obtaining the coefficient value from ASHRAE table [8] which is automatically assigned to each entity according to the fitting type.

Automatic fitting detection in Revit and finding the pressure loss coefficient in the ASHRAE table may seem like the best choice for the designer. But this method has some disadvantages. The first is the inability to edit the table (inability to edit the coefficient values and parameters) [9] and the second is the imperfect searching for duct fitting type. The table consists of geometric parameters of a defined duct fitting type, according to which Revit finds the appropriate coefficient value. The problem occurs when the geometric parameters of the duct fitting do not correspond exactly to the tabular values of the geometric parameters. In Czechia, duct fittings are often made-to-measure and that is the reason why there is no appropriate value in ASHRAE table. In these cases, the table is unable to interpolate intermediate parameter values. The exact type of fitting is often difficult to detect (for example, the difference between smooth radius elbow and mitered elbow).

The solution of these imperfections can be free access to tabular values and the possibility of editing them. The ability of interpolation is also needed or the utilizing of mathematical equations expressing coefficient values depending on geometric parameters and more accurate detection of duct fitting type.

4. Complex pressure loss solution in BIM

The data use from the building information model for the purpose of determining the ductwork pressure loss can be utilized by several ways. When using commercial software, information can often be accessed with an external application using API (Application Programming Interface). As a result, plenty of parameters that are part of the software information model can be utilized as input variables to the physical models created in the source code of the external application.

An alternative to creating an external application may be to process IFC file. IFC file is an interchangeable format containing the fundamental parameters of the information model. This file contains number of parameters that can be used to analyse duct fitting pressure loss. The advantage is the versatility of this file. Commercial BIM tools can work with this type of file. However, the disadvantage is the limitation in the number of parameters contained in the file. In particular, the physical parameters with the duct fitting type are missing.

In both cases, it is possible to use geometric parameters and process them further. The main part of the analytical tool is its physical model (physical equations), according to which the calculation runs. In the case of pressure loss, the key factor is local pressure loss, its values often vary in the literature. They are most often written in tabular form. These coefficient values are currently often obtained by numerical methods, but with some simplifications they may differ from the actual measured values on a real ductwork. Tabular values may be unsuitable for use in automated calculation, as their values need to be interpolated and there is a risk that the entered parameters will be out of the range of values in the table. In such a case, it seems more appropriate to use approximation equations describing the dependences of the magnitude of the coefficient of local pressure losses on the geometric parameters of the duct fitting. For many duct fittings, the approximation equations have not yet been determined or are determined based on numerical analyses, not real ductwork experiments.

5. Results and discussions

Laboratory experiment to determine pressure losses on the ductwork is time consuming but corresponds most to the real state of the ductwork in the building. From the values measured in this way, it is possible to compile equations, which can be used in the automation of pressure loss calculations in the building information model. An example of an automatic pressure loss calculation could be an external application / add-on for Revit that calculates pressure losses for fundamental duct fittings, which are elbows and transitions for round and rectangular pipe cross-sections. On the internet, this application called Revit Ductwork Pressure Loss Calculator is located on the GitHub web hosting in the form of an open-source add-on for the Revit application. It is an externally accessing application that uses the internal parameters of the Revit software to determine the type of duct fitting and then assigns the appropriate pressure loss coefficient to each fitting utilizing geometric parameters. The coefficient is assigned from two different sources. One source is a tabular value according to the ASHRAE table from 2001 and the other is mathematical equation reflecting the dependence of the coefficient on the geometry. For these purposes, the laboratory experiment research established analytical equations depending the coefficient of local pressure losses on 45° and 90° elbows geometry parameters. The equations are incorporated into the application and can be found in the Table 1.

The advantage of the application is the possibility of its user editing. Editing tabular values, including the values of their parameters, is user-friendly without the need to intervene in the source code, as these values are stored in a text file. It is therefore possible to adapt the coefficient values to designer experience or preferences. The mentioned application has the potential in Revit software to be developed into an automatic tool for the analysis of pressure losses for the Czech (and or international) environment. The application development consists mainly in adding other duct fittings to the source code, which the algorithm would consider and thus contain a satisfactory number of fittings over time. Other elements suitable for inclusion in the calculation could also be the effect of elbows that are placed downstream nearly each other with small or no duct straight part. This advanced analysis can lead to more accurate results of pressure losses. For two consecutive elbows, the tabular values are often unsatisfactory, and the analytical expression of the coefficient-geometry dependency is not common. For this purpose, rectangular cross-section elbows with different aspect ratios of cross-sections and different values of axial rounding and different sizes of the straight intermediate part between the elbows were analysed. The result is the equation that can be incorporated into an analytical tool that utilizes information in a building model. The equations of pressure loss dependence are given in the following Table 1.

Table 1. Dependency of pressure loss coefficient on the geometry (W is width of the rectangular duct cross-section, H is heigh of the rectangular duct cross-section, L is length of intermediate straight section).

Ductwork type	Dependency of pressure loss coefficient and duct fitting geometry
Elbow 1 × 45°	$\xi = -12{,}970\left(\frac{W}{R}\right)^2 + 31{,}350\frac{W}{R} - 18{,}490$
Elbow 1 × 90°	$\xi = -5{,}606\left(\frac{W}{R}\right)^2 + 14{,}210\frac{W}{R} - 8{,}470$
Elbows 2 × 45°	$\xi = 0{,}043\left(\frac{H}{W}\right)^{-2{,}441} - 0{,}015\left(\frac{L}{W}\right)^3 + 0{,}072\left(\frac{L}{W}\right)^2$
Elbows 2 × 90°	$\xi = 0{,}432\left(\frac{H}{W}\right)^{-1{,}248} - 0{,}007\left(\frac{L}{W}\right)^3 + 0{,}032\left(\frac{L}{W}\right)^2$

Table 2. Measurement errors, equation validity intervals, residual sum of squares and level of significance (W is width of the rectangular duct cross-section, H is heigh of the rectangular duct cross-section, L is length of intermediate straight section).

Ductwork type	Absolute error	Validity interval	Residual sum of squares (RSS)	Level of significance α
Elbow 1 × 45°	$a = -12{,}970 \pm 0{,}350$ $b = 31{,}350 \pm 0{,}830$ $c = -18{,}490 \pm 0{,}480$	$\frac{W}{R} = (1{,}0244; 1{,}3548)$	0,339	0,95
Elbow 1 × 90°	$a = -5{,}606 \pm 0{,}38$ $b = 14{,}210 \pm 0{,}900$ $c = -8{,}470 \pm 0{,}523$	$\frac{W}{R} = (1{,}0244; 1{,}3548)$	0,395	0,95
Elbows 2 × 45°	$a = 0{,}043 \pm 0{,}004$ $b = -2{,}441 \pm 0{,}068$ $c = -0{,}015 \pm 0{,}001$ $d = 0{,}072 \pm 0{,}005$	$\frac{H}{W} = (0{,}254; 1{,}000)$ $\frac{L}{W} = (1{,}476; 3{,}952)$	8,578	0,95
Elbows 2 × 90°	$a = 0{,}432 \pm 0{,}030$ $b = -1{,}248 \pm 0{,}043$ $c = -0{,}007 \pm 0{,}004$ $d = 0{,}032 \pm 0{,}016$	$\frac{H}{W} = (0{,}254; 1{,}000)$ $\frac{L}{W} = (1{,}476; 3{,}952)$	30,390	0,95

6. Conclusions

The building information model stores a substantial amount of information that can potentially be used. The more information the model contains, the more practical utilizing ways there are. Utilizing the BIM model for analytical tasks is an advanced feature of design software and offers considerable potential, which allows us to make competent decisions based on analysis and data at a time when we can influence the shape of the building with a minimum of costs. By simply using the already existing parameters and supplementing the physical calculating, it is possible to automate and to streamline building analysing. In this way, we can also maximize the use of available information in the model. In addition to the benefit rapid transfer of all created information about the building, BIM also offers the potential for application in the physical construction field.

The ductwork pressure losses determination is necessary for the design of ventilation system in a building. This process can be lengthy and complicated. The fundamental variables for the calculation, which are the coefficients, often differ and are not uniform for the Czech environment. In this respect, the unification of values or mathematical dependencies can be a step towards standardizing this analysis and at simplifying, refining, and speeding it up. With simple automation processes, we can save a considerable amount of time and thus design buildings efficiently in the context of current needs.

References

[1] P. Smith, "BIM implementation - global strategies," *Procedia engineering*, pp. 482-492, 2014.
[2] P. Schild, and M. Mysen, "Recommendations on Specific Fan Power (SFP) and Fan System Efficiency –IV," *TZB-info*, Prague: Jícha, 2014.
[3] J. Chyský, and K. Hemzal, "Ventilation and air conditioning," Third completely reworked edition. *Prague: B press*, Brno; 1993.
[4] I. Idelchik, D. Grunaer, and A. Barouch, "Handbook of Hydraulic Resistance," X/5. Moskva-Leningrad: *Gosudarstvennoe Energeticheskoe Izdatel'stvo*, 1960.
[5] K. Laboutka, and T. Suchánek, "Local pressure loss coefficient values – heat sources and fundamental duct fittings," *TZB-info*, Praha, ©2001-2019.
[6] J. Schwarzer, and V. Zmrhal, "Determination of local pressure loss coefficients utilizing," *Article*. 2010.
[7] S. Genić, I. Arandjelović, P. Kolendić, M. Jarić, N. Budimir, and V. Genić, "A Review of Explicit Approximations of Colebrook's Equation," *Review*, 2011.
[8] R. Howell, H. Sauer, and W. Coad, "Principles of Heating, Ventilating, and Air Conditioning," *Ilustrated edition*, Atlanta: American Society of Heating, Refrigerating and Air-Conditioning Engineers, 2005.
[9] Autodesk. It's not possible to modify ASHRAE Table in Revit. Autodesk Knowledge Network, Autodesk Inc.; 2017.

Problems and Solutions for Green Roofs

I N Maltseva[1], D M Shvalev[1], K A Tkachuk[2]

[1] Department of Architecture, Ural Federal University named after the First President of Russia B. N. Yeltsin, Mira st. 17, Yekaterinburg, 620002, Russia
[2] Institute of Energy Efficient and Sustainable Design and Building, Technical University of Munich, Arcisstraße 21, Munich, 80333, Germany

i.n.maltceva@urfu.ru

Abstract. Green roofs have been the standard design in many countries for hundreds of years due to the excellent thermal insulation properties of the fertile layer. A green roof is a space created by adding additional layers of fertile soil and plants over a traditional roofing system. Green roofing should not be confused with a traditional roof garden where plants are placed in free-standing containers on an exploited roof terrace. This article discusses the main advantages of roofing landscaping, the factors that influence the choice of type of roofing depending on climatic conditions and the main types of design solutions in megacities. The authors analyse the influence of the green roof on the energy efficiency of the building in the climatic conditions of Yekaterinburg. The main disadvantages of landscaped exploited coatings are considered. Also proposed is a new technology for the construction of an operated roof without a roll coating, but using penetrating waterproofing of reinforced concrete structures. It has now been proven that a properly designed and installed green roof can replace or compensate for the green environment at ground level that a large city is deprived of as a result of urbanization. The use of green roofs gives the metropolis ecological, social, economic and aesthetic benefits.

1. Introduction

Rapid urbanization as a global challenge arose several decades ago, and the transformation of large cities into metropolises of glass and concrete has led to significant environmental degradation and lack of green areas in urban environments.

The problem of creating environmentally comfort zones in an aggressive urban environment is particularly important due to the rapid growth of urban population and the compaction of urban development.

An important direction in the development of the urban architecture is the development of a modern concept for the formation of green comfort zones. Nowadays this problem is solved by means of traditional methods, i.e., creation of parks, squares, increase in greenery of city streets. However, one of the most promising solutions in this direction is the arrangement of green surfaces on the roofs of buildings, stylobates, and underground car parks, especially near courtyard areas.

In the project "Roof Gardens" (author N.P. Titova), as early as the beginning of the XXI century, considers the possibility of significant improvement of the ecological situation in large cities through the use of empty roof surfaces and other artificial bases for placing objects of landscape architecture [1]. V. A. Gorokhov was engaged in solving the problems of the ecological state of the modern urban environment and described the basic principles of the organization of green areas in urban development in his work "Green nature of the city" [2].

Buildings and structures with green operated roofs will stand out from the general city landscape and become not only one of the additional zones of ecological comfort, but also a genuine decoration of modern metropolises, where each square meter of free space counts (figure 1) [3-7].

Figure 1. Greening of roofs in Moscow [8].

Russia also has experience in green roof design [9]. For example, the green courtyard of the "Stone Brook" with bridges and staircases is located on the roof of the parking lot and the shopping gallery in Yekaterinburg (Figure 2).

Figure 2. The green courtyard of the "Stone Brook" with bridges and staircases is located on the roof of the parking lot and the shopping gallery, Yekaterinburg.

Smartly designed and installed green roof compensates for the natural environment at ground level, which is deprived of a large city due to urbanization [10]. The main advantages of operated green roofs are following:

- improvement of urban air quality (reduction of carbon dioxide concentration);
- reduction of peaks and total drainage flow volume (due to the ability of the green roofs to hold and store rainwater);

- reduction of energy consumption of the building, both in cold and summer;
- creation of habitats for animals (insects, birds);
- improvement of the aesthetic attractiveness of the building through the created environment.

2. Analysis of the green roofs' impact on buildings' energy efficiency

Green roofs are an effective tool for improving the energy efficiency of buildings in both hot and cold climate regions. In the first case, they protect the building against overheating, in the second case they become an additional layer of thermal insulation that retains heat inside the building.

Today, green roofs are sophisticated engineering systems. They include an inverted roof base (effective thermal insulation from EPPS slabs, roofing membrane). Components such as drainage layer, filtering geotextile, soil substrate with a special plant layer resistant to climatic influences are placed on top of the standard inverted structure [11].

Consider two coating options: the first is a non-exploitable inverted type of roofing, typical for a traditional residential building (Table 1), and the second is a monolithic roll-free coating with greening (Table 2).

Comparative heat engineering calculation for the climatic conditions of the city of Yekaterinburg is carried out. The required thermal resistance of the structure $R_0^{тр}$, based on regulatory requirements for the reduced heat transfer resistance, we find by formula 1 [12]:

$$R_0^{reg} = a \cdot GSOP + b = 0{,}0005 \cdot 5613{,}4 + 2{,}2 = 5{,}01 \frac{m^{2} \cdot {}^\circ C}{W} \quad (1)$$

where a and b - data ratios from [11, Table 3] for the relevant types of buildings (for coating: a=0,0005; b=2,2).

The degree-days of the heating period are determined by the formula 2 [12]:

$$GSOP = (t_{int} + t_{hp}) \cdot z_{hp} = (-20 - (-5{,}4)) \cdot 221 = 5613{,}4\ {}^\circ C \cdot day \quad (2)$$

The values for the length of heating period z_{hp} and the average air temperature t_{hp} are taken according to [13, Table 1].

Table 1. Non-exploitable inverted type of roofing.

№	Material	Thickness σ, mm	Thermal Conductivity Coefficient λ, W/m°C	Thermal resistance R, m^2·°C/W
1	Ballast (washed gravel)	-	-	-
2	Geotextile (heat treated)	-	-	-
3	Extrusion polystyrene foam Technonicol CARBON PROF t = 100+100=200 mm	200	0,032	6,25
4	Geotextile (needle punched)	-	-	-
5	Waterproofing Technoelast EPP (2 layers)	10	0,17	0,06
6	Reinforced cement-sand screed (on a slope, t_{min}=50 mm)	50	0,76	0,07
7	Monolithic reinforced concrete plate B25, F75, W4	200	1,92	0,10
	Total thermal resistance of coating №1			6,48

Resistance to heat transfer with successive uniform layers is determined as the sum of the thermal resistances of the individual layers (formulas 3 and 4), which is typical for first coating:

$$R_1^{con} = \frac{1}{\alpha_{int}} + \frac{\delta_1}{\lambda_1} + \frac{\delta_2}{\lambda_2} + \cdots + \frac{1}{\alpha_{ext}} = \frac{1}{8,7} + 6,48 + \frac{1}{23} = 6,63 \frac{m^2 \cdot {}^\circ C}{W} \quad (3)$$

Taking into account the heterogeneity of the structure:

$$R_1^{act} = R_1^{con} \cdot r = 6,63 \cdot 0,92 = 6,09 \frac{m^2 \cdot {}^\circ C}{W} > R_0^{reg} = 5,01 \frac{m^2 \cdot {}^\circ C}{W} \quad (4)$$

Table 2. Monolithic roll-free coating with greening.

№	Material	Thickness σ, mm	Thermal Conductivity Coefficient λ, W/m°C	Thermal resistance R, $m^2 \cdot {}^\circ C/W$
1	Soil with greenery	200	1,05	0,19
2	Profile membrane Planter	8,5	-	-
3	Geotextile (heat treated)	-	-	-
4	Extrusion polystyrene foam Technonicol CARBON PROF t = 100+100=200mm	200	0,032	6,25
5	Geotextile (needle punched)	-	-	-
6	Waterproofing Technoelast GRIN (1 layer)	10	0,17	0,06
7	Waterproofing Technoelast EPP (1 layer)			
8	Reinforced cement-sand screed (on a slope, t_{min}=50 mm)	50	0,76	0,07
9	Monolithic reinforced concrete plate B25, F75, W4	200	1,92	0,10
		Total thermal resistance of coating № 2		6,67

Resistance to heat transfer for the second coating is calculated by formulas 5 and 6.

$$R_2^{con} = \frac{1}{\alpha_{int}} + \frac{\delta_1}{\lambda_1} + \frac{\delta_2}{\lambda_2} + \ldots + \frac{1}{\alpha_{ext}} = \frac{1}{8,7} + 6,67 + \frac{1}{23} = 6,83 \frac{m^2 \cdot {}^\circ C}{W} \quad (5)$$

Taking into account the heterogeneity of the structure:

$$R_2^{act} = R_2^{con} \cdot r = 6,83 \cdot 0,92 = 6,28 \frac{m^2 \cdot {}^\circ C}{W} > R_0^{reg} = 5,01 \frac{m \cdot {}^\circ C}{W} \quad (6)$$

Difference of reduced heat transfer resistances

$$\Delta R = 0,19 \frac{m^2 \cdot {}^\circ C}{W} \quad (7)$$

$$R_1^{con} = 6,09 \frac{m^2 \cdot {}^\circ C}{W} < R_2^{act} = 6,28 \frac{m^2 \cdot {}^\circ C}{W} \quad (8)$$

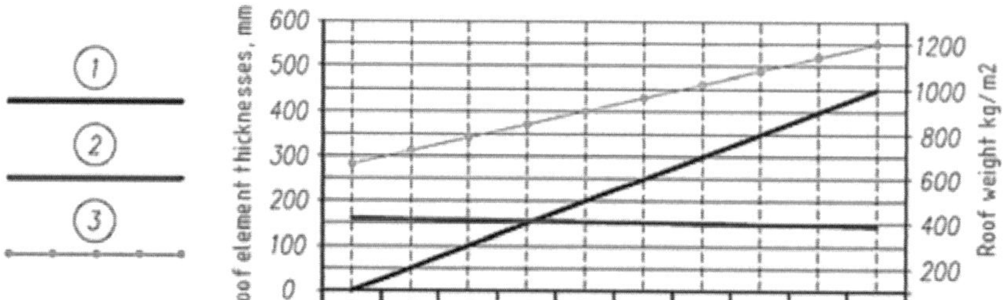

		1	2	3	4	5	6	7	8	9	10
1	Vegetation layer thickness, mm	0	50	100	150	200	250	300	350	400	450
2	EPPS insulation thickness, mm	160	158	157	155	154	152	151	149	148	146
3	Roof weight kg/m2	665	725	785	845	905	965	1025	1085	1145	1205

Figure 3. The graph of the dependence of the weight of the roof on the thickness of the vegetation layer and insulation, with a constant thermal resistance of the enclosing structure 5,2 $m^2 \cdot °K/W$ (Coating 2).

Based on the results obtained (figure 3), we can conclude that the most optimal type of roof greening in the climatic conditions of Yekaterinburg is the extensive type with a soil substrate thickness of 150-200 mm.

This technology combines: improving the thermal performance of the roof, protecting the building from overheating in the summer with a relatively small weighting of the roof – 100-150 kg/m^2.

3. Features of the arrangement of green operated coatings

In contrast to the usual flat roof, which is calculated only on the impact of climatic loads, the exploited one is exposed to increased mechanical loads arising from crowding, traffic, loads from the own weight of materials used to create functional space.

The additional load of the green surface depends on the nomenclature of plants used, the weight of the soil and gravel layer in the wet state, the weight of the additional road surface, the weight of the necessary equipment, etc. (Figure 4) [14].

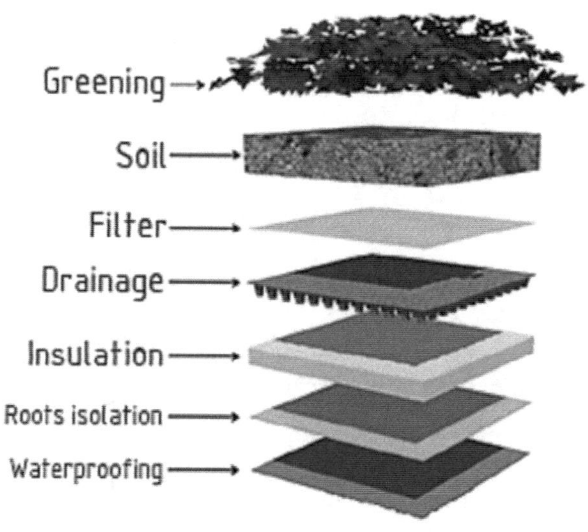

Figure 4. Schematic structure of the "green roof" [15].

There are two main types of green roofing systems: *extensive* and *intensive*. They differ mainly in the choice of plants used, the thickness of the plant environment, as well as cost.

Extensive green roofing systems characterized by:

- low weight (plant layer 0,10-0,20 m, wet weight 100-250 kg/m^2);
- low variety of used plants (plants with high resistance to climatic influences)
- minimum roof maintenance.

Intensive green roofing systems characterized by:

- high weight (plant layer 0,20-0,60 m, wet weight 250-950 kg/m^2);
- high variety of used plants;
- high requirements for roof maintenance (need to use special drainage and irrigation systems).

Traditional green operated roofs are characterized by extremely low repairability, so the choice of components, and especially the arrangement of the waterproofing layer must be approached with great responsibility.

If the waterproofing layer of the used roofing starts to leak over time, it will be necessary to remove the roof structure above it. Therefore, both the bearing elements and the waterproofing materials of the green roofing must meet the highest requirements for strength and durability.

4. Roll-free type of roofing

The waterproofing layer protects the structural elements of the roof from the harmful effects of external climatic factors, such as moisture, frost, ultraviolet, etc. Most often the service life of the whole building depends on the reliability of the waterproofing layer.

Traditionally, roll-fed or coated materials are used for waterproofing concrete structures. They protect the concrete from moisture, but work separately from the main structure. During operation, such materials peel off the base, break down and fully stop working, resulting in leaks [16].

To address the problems of durability and quality of roof coverings, EPK "Dokros" patented a new type of roof "Reinforced Concrete roll-free roofing» (Figure 5).

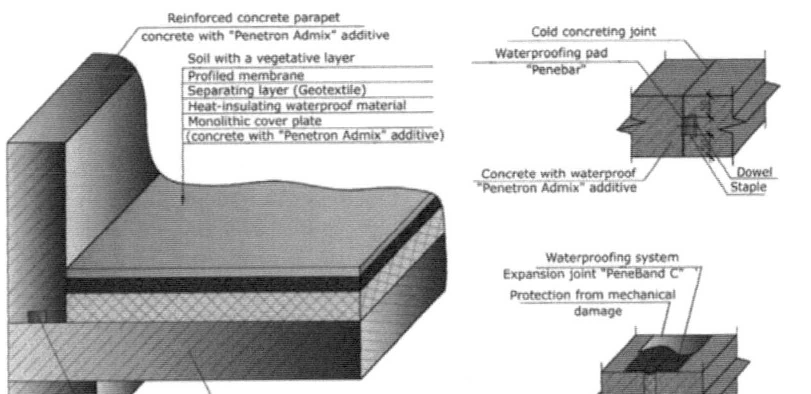

Figure 5. Simplified scheme of a roll-free roofing system.

The roof structure consists of monolithic reinforced concrete slabs (plates) separated by expansion joints depending on the size of the covering. The structural parameters of the covering slabs (section height, reinforcement, concrete class) are selected based on the working conditions and estimated load on the roof [17].

The necessary properties of the concrete in terms of density and waterproofing are ensured by selecting the appropriate additives from the "Penetron" system for the initial waterproofing, which eliminates the need of using traditional roofing materials with short service life.

It is enough to waterproof only the concrete slab using the special additive for concrete "Penetron Admix", which will ensure a high level of waterproofing of the structure already at the construction stage [18]. Waterproofing of concreting joints, as well as places of entry of communications, is carried out with materials of the "Penebar" system.

The main advantages of a roll-free type reinforced concrete roof:

- increasing the reliability and durability of the roofing structure (the service life of the initial waterproofing is equal to the service life of reinforced concrete up to 100 years);
- reduction of labor costs for roofing (exclusion of secondary waterproofing);
- weight reduction of roofing by 10-15 kg/m^2.

In addition, the components of the "Penetron" system make it possible to repair the roof from the inside of the room, which significantly reduces the complexity, and most importantly, the cost of repair work, especially on green operated roofs (Figure 6) [19].

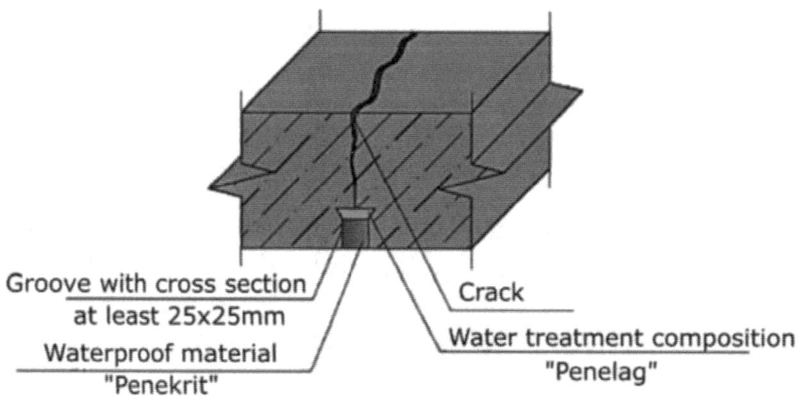

Figure 6. Elimination of pressure leaks and sealing cracks in a monolithic reinforced concrete roof from inside the room.

This roof will serve as a reliable basis for the setup of a green operated roofs, both on the high-rise part of the building, and above the parking lot or stylobate of any architectural form.

5. Conclusion

The implementation of operated green roofs can solve many problems, including reduction of the number of grey, unremarkable roofs of existing buildings, as well as solving the problem of the lack of urban spaces for creating zones of environmental comfort.

However, an important engineering task is to increase the reliability and durability of green operated roofs, due to the expensive maintenance and repair of traditional roofing systems using roll-fed waterproofing materials.

One of the solutions to this problem is using green operated roof coating based on a roll-free monolithic coating, which not only simplifies the installation of the coating, but also reduces the cost of its operation. In addition, the absence of a fragile waterproofing layer significantly expands the possibilities in the choice of systems and technologies for the operated roof setup.

References

[1] N. P. Titova, "Roof gardens," *Moscow: OLMA-PRESS Grand* (In Russian), pp. 58–77, 2002.
[2] V. A. Gorochov, "Green nature of the city," *Moscow: Stroyizdat* (In Russian), pp. 24-46, 2003.
[3] "Supermarket Rooftop Kindergarten, Melbourne, Australia," *J. Ecosoil* (In Russian), accesible from: https://www.ecosoil.ru/useful/luchshie-sady-na-kryshah/, 2015.
[4] "Roof garden of the MOM Park Shopping Center parking lot, Budapest, Hungary," *J. Ecosoil,* (In Russian), accesible from: https://www.ecosoil.ru/useful /luchshie-sady-na-kryshah/, 2020.
[5] "Children's Hospital, Philadelphia, USA," accesible from: https://architizer.com/projects /childrens-hospital-of-philadelphia/, 2015.
[6] "Used roof of a modern parking lot - a project from Sweden," *J. Green roofs. Design and construction* (In Russian), accesible from: https://www.grrus.ru/news/ispolzuemaya-krovlya-sovremennogo-parkinga-proekt-iz-shvecii/, 2020.
[7] "Green Roof in Chicago - a masterpiece of modern eco-architecture," *J. Culturologia (In Russian),* accesible from: https://kulturologia.ru/, 2018
[8] Roof Greening (plant species) (In Russian), accesible from: https://zen.yandex.ru/media/rieltor/ozelenenie-krovli-vidy-rastenii-5db3dfde9c944600b0e7b44c, 2019.
[9] "New trend of new buildings: overview of houses with a "green roof" in Yekaterinburg," 2017.
[10] T.N. Belikova, "Green roofs or "Gardens of Semiramis"," *J. Complex problems of the development of science, education and economy of the region*, vol. 2(5), pp. 18-29, 2014.
[11] "Green roofs in cold climates," *J. Vogean* (In Russian), accesible from: http://vogean.com/katalog/538/index.php, 2017.
[12] SP 50.13330.2012 Thermal protection of buildings, *Moscow: Stroyizdat (In Russian),* 2012.
[13] SP 131.13330.2012 Building climatology, *Moscow: Stroyizdat (In Russian),* 2012.
[14] "Recommendations for the design of landscaping and improvement of roofs of residential and public buildings and other artificial foundations," *Moscow: Stroyizdat (In Russian),* pp. 27-28, 2000.
[15] Kryshikrovli. Accesible from: http://kryshikrovli.ru.
[16] "Waterproofing of floor slabs and the operated roof of the above-ground parking," *In Rus: Internet J. Sukhoy zakon,* accesible from: http://www.s-zakon.ru/catalog/popyliarnoe-penetronovedenie/gidroizoljacija-plit-perekrytija-i-ekspluatiruemoj-krovli-nadzemnogo-parkinga.html, vol.5 (132), 2017.
[17] V.V. Domnin, and K.V. Domnin, "Monolithic reinforced concrete rollless roofs for residential and public buildings. (Technical solutions)," *Yekaterinburg: EPK LLC "Dopros"* (In Russian), *2011.*
[18] V.V. Domnin, and K.V. Domnin, Utility model patent No. 99037 Russian Federation "Reinforced concrete roof structure and method of its repair," 2019.
[19] STO-6658209531-003-2017 Repair of concrete and reinforced concrete lining of transport tunnels and underground structures of subways using materials from JSC "GK Penetron-Russia".

IOP Conf. Series: Materials Science and Engineering 1066 (2021) 012013 doi:10.1088/1757-899X/1066/1/012013

Quality of the Indoor Environment in the Nursery School: Monitoring, Analysis and Optimization

Michal Kraus [1], Ingrid Juhásová Šenitková [1], Kamila Hadrbolcová [1]

[1] Institute of Technology and Business in České Budějovice, Department of Civil Engineering, Okružní 517/10, 370 01 České Budějovice, Czech Republic

19514@mail.vstecb.cz

Abstract. The paper aims are to assess and evaluate the hygiene and quality of the indoor environment in selected indoor areas of the kindergarten, to analyse the construction, technical and technological solutions of the building in accordance with the concept of designing healthy buildings and their environment. The paper will propose the operation of the building in terms of hygiene and perceived air quality. The paper assumes in-situ monitoring of the indoor environment of the nursery school using a multifunctional measuring instrument for analysing the quality of the indoor environment and comfort (Testo 480 + accessories). The research goal is to update and expand information on the quality of the indoor environment of kindergartens during the heating season. The aim is to design a concept for improving the quality of the environment in nursery schools and to prepare documents for relevant recommendations, including recommendations concerning, for example, construction or utility materials for nursery schools and the mode of their operation.

1. Introduction

The lifestyle of a large part of the European population is focused on spending most of the day indoors [1]. At present, the topic of the quality of the indoor environment of buildings is an increasingly discussed issue. The importance of the quality of the indoor environment and especially hygiene is growing in connection with the global epidemiological situation regarding coronavirus disease SAR-CoV-2 [2-4]. Increasing attention is paid to the quality of the indoor environment in schools, and therefore in nursery schools. The main reason for focusing on school's environment is the fact that children are gathered in large numbers in closed rooms, in which the overall air quality is often forgotten for reasons of thermal comfort. School classes tend to have up to 3-4 times more people than in office spaces (with regard to floor area m^2 and air volume m^3). Another reason for the increased interest is the fact that children, especially the youngest, are considered to be the most vulnerable group that can be most harmed by poor air quality [5-7]. Generally, harm to the child's body is greater due to the inhalation of a larger amount of air per unit body weight compared to an adult. The quality of the indoor environment also affects the perception of comfort or the feeling of health and well-being. At the same time, it creates an environment that can have both positive and negative effects on the study environment for children and pupils, but also teacher performance [8].

The relevant legislative regulations establish requirements for spatial conditions, equipment, operation, lighting, heating, climatic conditions, water supply, cleaning and kindergartens and other school equipment and even for children younger than 3 years. The interior of the preschool classroom must include space for instruction, free play, rest, personal hygiene with hardening, physical exercise and catering (possibly in another canteen, such as a school canteen nearby). Premises intended for physical education must be ventilated, the floor must be made of flexible material, the surfaces must be washable, and the radiators must be secured against injury to the child.

2. Experimental method

The researched nursery school building is designed from a modern modular cellular system. The cells are assembled into two floors so that the first floor is for the youngest children, approximately 2 to 4 years of age, and the second for preschool children, approximately 4 to 6 years of age.

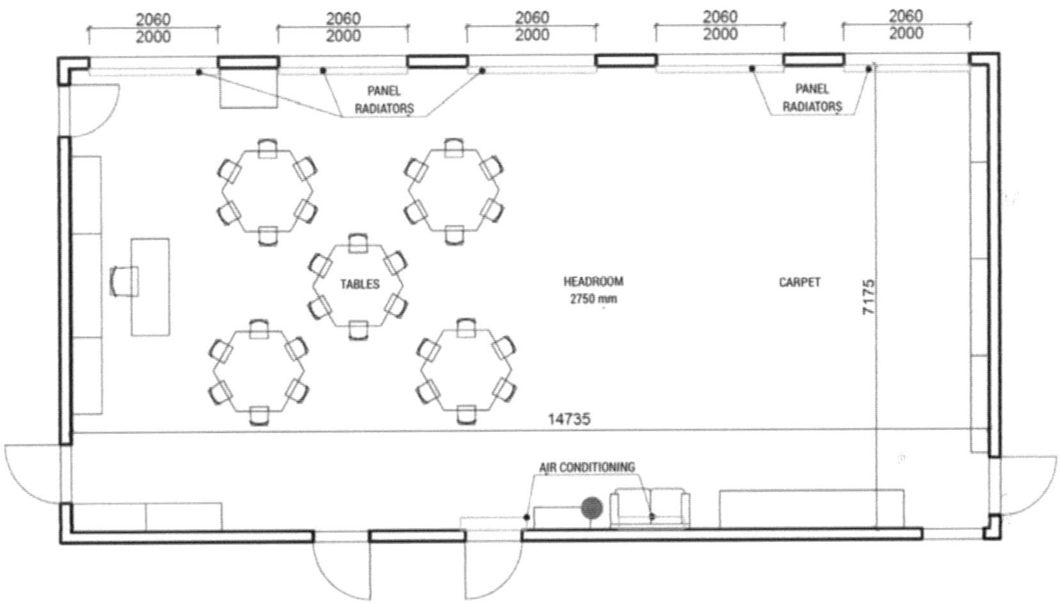

Figure 1. Scheme of the monitored class of nursery school.

Figure 1 schematically shows the arrangement of the monitored class. The red dot shows the position in which the measuring instrument was placed throughout the measurement. According to the children's activity, the sensor sensing the data was moved towards their current position. Children spend most of their time at tables where food is served or where children learn and play. Children spend less time on the carpet, where games, learning and exercise take place in both classes.

The monitored class is rectangular with dimensions of 14,735 x 7,175 mm with a clear height of 2,750 m. The total volume of the class is 290.74 m^3. There are 5 window openings in the class with dimensions of 2,060 x 2,000 mm filled with three types of window assemblies. These are assemblies of two windows, of which one window is always openable and the other window, placed vertically or horizontally, is tiltable. The plate radiator is under each of the window. The dimensions of which reach a maximum height of the windowsill, which is 450 mm high from the floor. Above the door to the washroom and about 3 meters further, there are two air conditioning units. The class is also equipped with plants. In the classroom are two potted plants placed on higher shelves.

The multifunctional measuring instrument Testo 480 type K was used to measure climatic variables in the classes of the selected kindergarten. In this work, the concentration of carbon dioxide [ppm], temperature [°C], humidity [%] and lighting intensity [lx] were measured. The IAQ probe was used to measure carbon dioxide concentration, temperature and humidity, which is used to quickly control the indoor climate. As stated by the manufacturer, the range of the temperature probe is from 0 to +50°C with an accuracy of ± 0.5°C. The range of the humidity capacitive sensor is from 0 to 100% with an accuracy of ± (1.8% + 0.7% of the measured humidity). And for CO_2 measurements, the scale is in the range from 0 to +10,000 ppm, with an accuracy for values from 0 to +5,000 ppm ± (75 ppm + 3% of the measured value) and values from 5,001 to +10,000 ppm ± (150 ppm + 5% of measured value). The second probe used is a probe for measuring light intensity [lx]. The measuring range of the probe used is from 0 to +100,000 lux. The accuracy of the probe is according to DIN EN 13032-1 for f_1 = 6% V-Lambda and f_2 = 5% cos, and at the same time it is classified in class C according to DIN 5032-7.

The measurement site should be chosen in close proximity to the building users for the most accurate results and at the height at which they breathe. The respiratory zone of children is located approximately 1 m above the ground, this height of the device was observed. However, for safety reasons, the meter accessories were placed at the wall of the room, as shown in Figure 1 above, at a close distance from the users of the building.

3. Results

Measurements and examinations in the classroom took place in the winter. The outdoor daytime temperature on these days ranged from -4 to 8°C. The daily schedule of the children did not change much, but the number of children present changed considerably. The device was switched on every day at 7 o'clock and 15 minutes. In the resulting graphs, however, only the values measured at 7 hours and 25 minutes are recorded in order to acclimatize the device and thus prevent errors in the initial measurements. To compare the values from the monitored days, the values are entered into individual graphs. In each of the following graphs, the values of one indicator during the whole week are recorded. Figure 2 shows occupancy by operating time. At 10 o'clock the children regularly went for a walk outside.

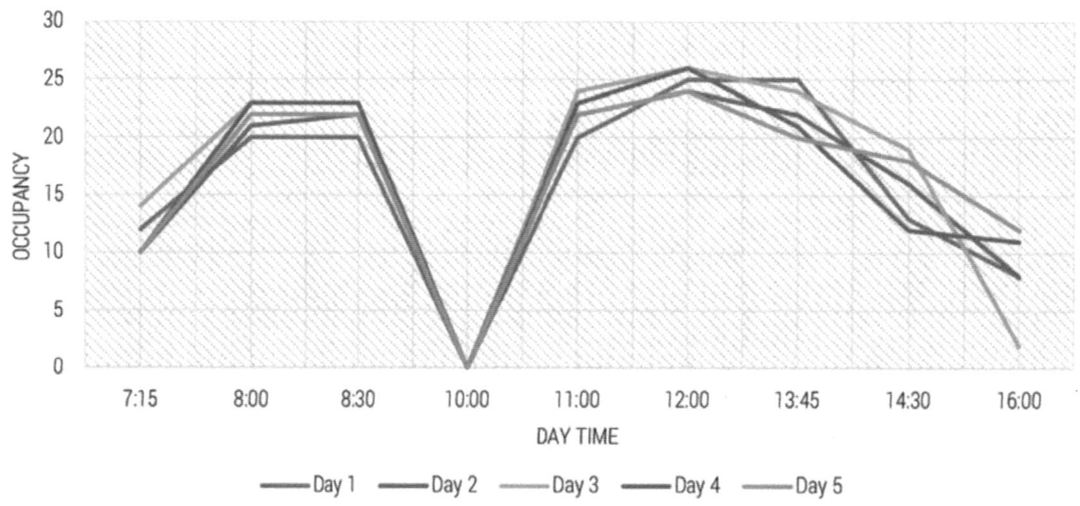

Figure 2. Occupancy during the operation day.

For an overview of the concentration of carbon dioxide in the classroom during the entire measurement, the data are entered in the following figure. The figure shows the course of CO_2

concentration each day, and this course can be compared with the value indicating the maximum allowable concentration. The maximum permissible limit of carbon dioxide concentration was exceeded only on the second day of the measurement. The measured space became an unsuitable environment in the case of the fourth day of measurement, when the limit of 1,200 ppm was exceeded twice. The limits were exceeded due to low air volume exchange caused by the absence of ventilation. CO_2 concentrations are shown on Figure 3.

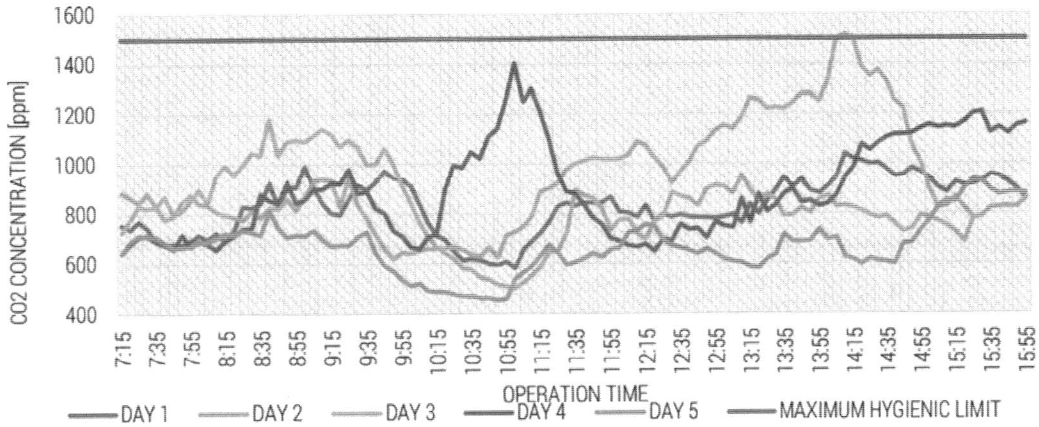

Figure 3. CO_2 Concentration [ppm].

The following graph (Figure 4) clearly shows the light intensity [lx] during operating hours. The value indicating the minimum permissible value (80 lx) for a well-lit space is also shown here. During the third day, the lighting intensity was exceeded over 4000 lx, which most likely caused dazzling users. Such a high value was measured due to direct irradiation of the sensor at the time the value was recorded. During the measurement, the light intensity fluctuated considerably due to the variability of solar radiation. The measured values were almost every day in lower or higher values than the ideal, which was caused by fluctuations in solar radiation.

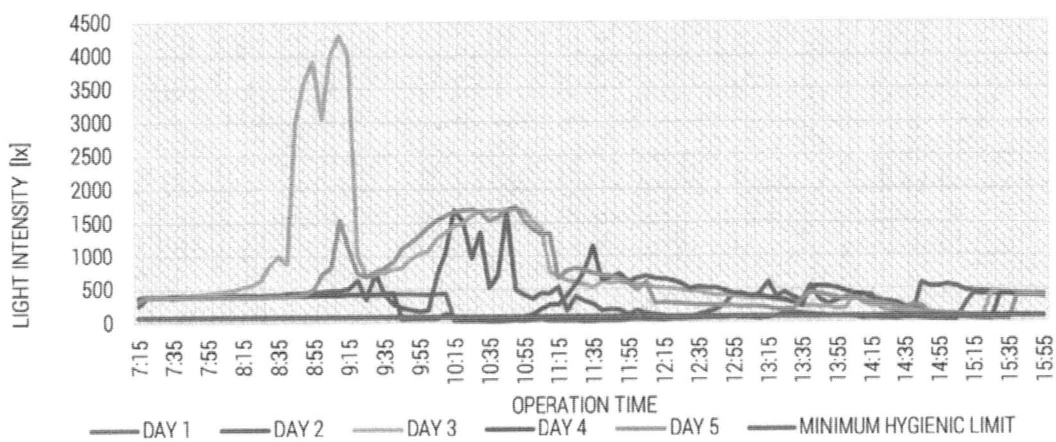

Figure 4. Light intensity [lx].

Figure 5 shows clearly the course of the relative humidity (RH) of each day. A value indicating the minimum permissible relative humidity (30%) is shown. Values below the minimum permissible relative humidity were monitored each day of the measurement.

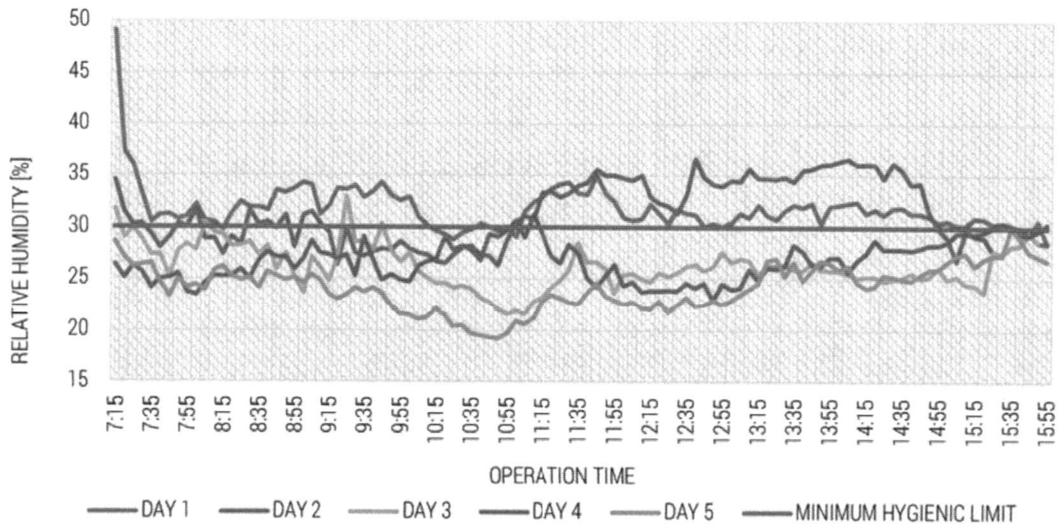

Figure 5. Indoor relative humidity [%].

Indoor air temperature is shown in Figure 6. The values of maximum and minimum permissible temperatures of air are shown (boundaries). The maximum or minimum value of the permissible air temperature was not exceeded during the whole week. Most of the time, even the measured values were in the ideal range of 20 to 24 °C.

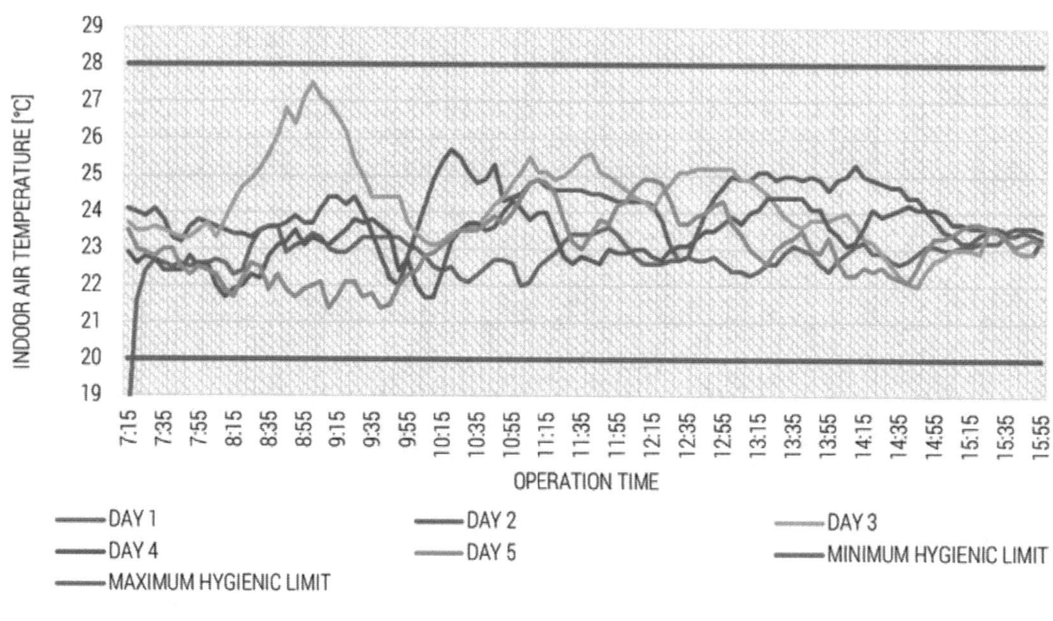

Figure 6. Indoor air temperature [°C].

4. Discussions

Table 1 shows a summary of the minimum, maximum and mean values in the measurement of the quality of the indoor environment determined by the nursery school. Color-coded values indicate that the recommended minimum and maximum values for the indoor environment have been exceeded.

The set limit of carbon dioxide concentration of 1500 ppm was exceeded only once in the class and once the values approached the limit. Also, in the monitored class, the CO_2 concentration did not fall below 1000 ppm, which is the limit value for an ideal environment. In terms of average values, the results correspond to ideal conditions.

The intensity of lighting fell several times below the limit of a well-lit room. In the first class even 4 out of five cases. According to the observations, these were mostly sunny days with significant fluctuations in the intensity of sunlight, or the switching off of artificial lighting in the absence of children. The average values correspond to the ideal light intensity, which should be between 200 and 300 lx.

In the case of relative humidity, values are indicated where the humidity is lower than 30% (too dry) or, conversely, higher than 60% (too humid). According to the table, it is clear that the monitored class is facing a dry environment problem. Minimum relative humidity values below 30% were recorded on all days.

The table shows that during the 4 days of measurement, values higher than values in the ideal range from 20 to 24°C were recorded. On average, however, the values were increased compared to the ideal only in two days, at a temperature higher by a maximum of 0.3°C.

Table 1. Results of monitoring the quality of the indoor environment.

	CO_2 concentration *[ppm]*			*Light intensity [lx]*			*Indoor relative humidity [%]*			*Indoor air temperature [°C]*		
	Min	*Max*	Ø	Min	*Max*	Ø	Min	*Max*	Ø	Min	*Max*	Ø
Day 1	581	1040	829	13	429	202	28.5	36.0	31.5	22.0	23.6	22.9
Day 2	618	1518	944	50	1684	421	26.3	36.6	31.3	22.0	25.7	24.1
Day 3	502	1181	831	81	4299	839	21.6	32.9	26.0	22.0	27.5	24.3
Day 4	648	1402	897	31	1141	351	22.9	31.2	26.8	21.7	24.4	23.3
Day 5	454	935	667	31	1742	591	19.2	30.0	24.1	21.4	24.9	23.1

The indoor air relative humidity is the biggest problem in the quality of the monitored nursery school. The relative humidity of indoor air was below the limit values for the vast majority of the measured time, and therefore it is necessary to focus primarily on increasing its humidity, depending on the indoor air temperature. Ensuring an optimal heat-humidity microclimate is one of the most important factors needed to achieve a healthy environment and indoor air. It is necessary to set a temperature that will not cause the air to dry out, as is the case with overheating. The higher the indoor air temperature, the drier the air becomes. It is necessary to ventilate effectively. The most ideal way is short but intensive ventilation. During the monitoring of the class, it was observed that the ventilation is performed on the contrary, weakly and continuously. Most of the time the measurement took place, one to four windows were opened for ventilation at the same time as the heating was switched on. This method of air exchange takes place in a cycle that is inefficient for indoor air quality. Cold air enters the open window, causing an unpleasant cold environment, which causes an attempt to heat the environment by heating, but a large part of the heated air escapes through the window from the room. The result is a dry environment with high energy consumption spent on equalizing the room temperature. Due to the higher recorded temperatures than the ideal values, with at the same time low values of relative air humidity, it should be a proposal to optimize the efficiency of the ventilation method at the same time as adjusting the heating intensity. Ventilation should take place at least 3 times a day for 10 to 15 minutes. In the winter months, ideally reduce the ventilation time to two minutes and turn off the heating a reasonable time in advance (approximately 30 minutes) and switch it on again after ventilation. If the relative humidity is still low, it would be appropriate to combine the air exchange method with other elements that increase the humidity, i.e. steam humidifier, ultrasonic humidifier, natural evaporation humidifier, evaporator for radiators or plants.

Improvement of ambient air conditions in the context of CO_2 concentration can only be achieved by supplying fresh air. According to the methodological guidelines for the design of school ventilation, the amount of air supplied to the classrooms can be reduced to 10 m^3/h per child due to the age of the children. The maximum capacity of the class is 28 children, which is 280 m^3 of air needed in a one-hour exchange. There are one or two teachers in the class, for whom the volume required for air exchange is 20 - 30 m^3/h per person. Together, for all persons, at least 320 m/h of the total volume of the class of 290.74 m must be replaced. It follows that when the maximum capacity of the class is filled, a complete air change will be required more often than once an hour.

During the measurement, the light intensity fluctuated considerably due to the variability of solar radiation. The measured values were almost every day in lower or higher values than the ideal, which was caused by fluctuations in solar radiation. The windows are equipped with blinds, but except for one case, they were not used for the whole ten days, despite the considerable dazzling of children at the moment when they were learning or forming at tables. At the same time, in at least seven measured days, the light intensity dropped to a limit of 80 lx for several tens of minutes and even exceeded this limit a few times. In the case of such decreases in light intensity, there was considerable gloom in the room, and yet no artificial lighting was used. Both extremes can be avoided by using blinds with the slats tilted downwards so that the sun illuminates the room but does not dazzle the children, and by promoting artificial lighting. At the same time, the blinds can be used to reduce any unwanted heat gain by turning the slats upwards.

5. Conclusions

The quality of the indoor environment is a great phenomenon today. Quality and healthy indoor environment is very often at odds with the current trend of low energy consumption. Nursery schools s are sensitive places for children due to the exposure of a large number of factors to the indoor environment of buildings. The aim is to design a concept for improving the quality of the environment in nursery school and to prepare the concept for relevant recommendations, including recommendations concerning, for example, construction or utility materials for nursery schools and the mode of their operation.

Acknowledgment(s)

The authors gratefully acknowledge the financial support received from the Institute of Technology and Business, under project no SVV202006. The authors also wish to thank local founder for support and access to the nursery school for these experiments and monitoring.

References

[1] C. Mandin, M. Trantallidi, A. Cattaneo, B. Canha, V: G. Mihucz, T. Szigeti, R. Mabilia, E. Perreca, A: Spinazze, S. Fossati, Y. D. Kluizenaar, E. Cornelissen, I. Sakellaris, D. Saraga, O. Hänninen, E: D. O. Fernandes, G. Ventura P. Wolkoff, P. Carrer, J. Bartzis, "Assessment of indoor air quality in office buildings across Europe – The OFFICAIR study," *Science of the Total Environment*, vol. 579, pp. 60-178, 2017.

[2] M. Y. Z. Abouleish, "Indoor Air Quality and Coronavirus Disease (COVID-19)," *Public Health*, in press.

[3] J. Saha, P. Chouhan, "Indoor air pollution (IAP) and pre-existing morbidities among under-5 children in India: are risk factors of coronavirus disease (COVID-19)?" *Environmental Pollution*, vol. 255, part 2, 115250, 2020.
[4] N. Clements, M. J. Binnicker, V. L. E. Roger, "Indoor Environment and Viral Infections," *Mayo Clinic Proceedings*, vol. 95, pp. 1581-1583, 2020.
[5] A. Stametelopoulou, D. N. Asimakopoulos, Tm. Maggos, "Effects of PM, TVOCs and comfort parameters on indoor air quality of residences with young children," *Building Environment*, vol. 150, pp. 233-244, 2019.
[6] S. Kim, J. A. Senick, G. Mainelis, "Sensing the invisible: Understanding the perception of indoor air quality among children in low-income families," *International Journal of Child-Computer Interaction*, vol. 19, pp.79-88, 2019.
[7] E. Midouhas, T. Kokosi, E. Flouri, "Outdoor and indoor air quality and cognitive ability in young children," *Environmental Research*, vol. 161, pp. 312-328, 2018.
[8] M. Kraus, P. Novakova, "Assessment of the Indoor Environment for Education," *IOP Conference Series: Earth and Environmental Science*, vol. 290, issue 1, 012144, 2019.

Problems of People's Evacuation Modelling

A Parfyonenko [1], J Kurbatova [1]

[1] National Research Moscow State University of Civil Engineering, Yaroslavskoye Shosse, 26, Moscow, 129337, Russian Federation

ikbs@mgsu.ru, kurbatovajuliapb@mail.ru

Abstract. In the field of the study of the human flows traffic, there are programs appear whose developers don't have the knowledge required about fire safety and the parameters of the human flows traffic. Algorithms embedded in software and computer systems give inadequate results, which reduces the indicator of people's safety in buildings and structures. This work describes the features of using individual-flow evacuation modeling of people in buildings and structures when changing the width of the aperture of door. The main parameters of the human flow, characteristic of individual-flow modeling, are also considered. The survey of software and computer systems used in risks estimation of buildings and structures. The results of people's evacuation modeling in software and computer systems are analyzed. An assessment of the correctness of the results was made, a conclusion was drawn on the adequacy and possibility of using these software and computer systems in calculating the time of evacuation of people in buildings and structures.

1. Introduction

Evaluation of planning decisions from the point of view of the safety of people's lives [1] in case of fire in buildings and structures inherently includes the calculation of evacuation of people. This article envisages the modeling of evacuation of people in various computer software and computing systems. In the field of the study of the human flows traffic, programs are emerging for simulating the evacuation of people, which, as practice shows [2-4], give inadequate results due to a lack of understanding by the developers of the patterns of human flows traffic. Therefore, the algorithms embedded in the programs must be checked for the correctness of the data obtained, and the programs themselves for the consistency of the laws used between the parameters of human flows and the real processes of movement of people. Therefore, the analysis of the developed computing software and computer systems, which are used to estimate the evacuation time, is an urgent task, and there is an urgent need to test the software systems and obtain appropriate conclusions from competent specialists before entering the public market. The modeling of the human flows traffic was considered in the works of Parfyonenko A "Methodology for modeling human flows and the practice of programming their movements during evacuation" [5] and Kholshchevnikova "Once again on the modeling of human flows in assessing fire risk" [6].

2. Methods

The Methodology for determining the calculated values of fire risk in buildings [7] presents three computational models of human flows traffic: simplified-analytical, simulation-stochastic and individual-flow models.

Simplified-analytical model allows us to establish calculated correspondence between the parameters of the human flows traffic [8,9]. This model is the simplest and most tried and tested. It is implied that people during evacuation create equable distributed flows, the mixture of evacuated flows (mobility group, projection area, speed) is not taken into account. Therefore, this model gives adequate results only when calculating the evacuation from geometrically simple units.

The simulation-stochastic model of the human flows traffic makes it possible to accurately assess the qualitative and quantitative picture of the human flows traffic during evacuation. When calculating, the building is divided into elementary sections, for each of which a lot of settlement operations are performed per second. In this model, there are phenomena peculiar to the human flow [10-12]: spreading, reformation, merging, decompression. Both models take into account the concept of human flow, are sensitive to changes in the geometry of the building, but don't represent the possibility of considering the traffic of an individual person in the flow.

The individual-flow model involves modeling the traffic of an individual person taking into account its individual parameters (speed, projection area, mobility). This model is used to calculate evacuation from a building with a complex internal layout [13].

The mathematical model is selected based on the features of the layout of the building, the area of horizontal projections of people, as well as the parameters of people's traffic of various mobility groups.

For the analysis of software and computing systems, design schemes for the people's traffic from a room 20x2 m in size with aperture of doors of 0,8, 1,2, 1,6 m wide and traffic from a room without an aperture of doors, i.e., on a horizontal section of the path 2 m wide. For the calculation, it was necessary to place 360 people in a room with a horizontal projection area of each person $f =$ 0,1 $м^2$/person and a density of people's flow $D = 0{,}9\ м^2/м^2$. According to the calculation schemes, the calculation results were obtained in accordance with the Methodology for determining the calculated values of fire risk [7]. The calculation of time was carried out according to the laws of human flows traffic:

$$t_{ск} = \frac{N \cdot f}{q_{при\ D=0,9} \cdot b} \tag{1}$$

where N – is the number of people, person; f – horizontal projection area of a person, $м^2$/person; $q_{при\ D=0,9}$ – the intensity of movement through the site at a density of 0,9 or more, m / min; b – the width of the section, upon entering which, a crowd is formed, m.

The intensity of traffic through the aperture of door with a width of less than 1,6 m is determined according to the dependence:

$$q = 2{,}5 + 3{,}75 \cdot x \tag{2}$$

where$x$ – is the width of the aperture of door, m.

Further, it was necessary to simulate the traffic of people through various apertures in software and computing systems and compare the results of manual counting for compliance with the Methodology [7]. It was assumed that if software and computing systems take into account the laws of traffic of people in accordance with the Methodology, then the results should be close to the values obtained in the calculation according to a simplified analytical model. If the deviation of the results is significant, then the use of these programs is unacceptable. Visualizations of modeling the human flows traffic are presented in Figures 1- 5.

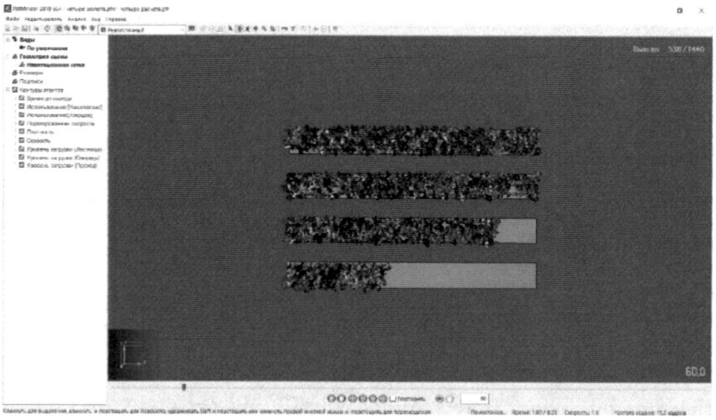

Figure 1. Pathfinder: visualization of the simulation of the human flow traffic through apertures of various widths.

Figure 2. Sigma PB: visualization of the simulation of the human flow traffic through apertures of various widths.

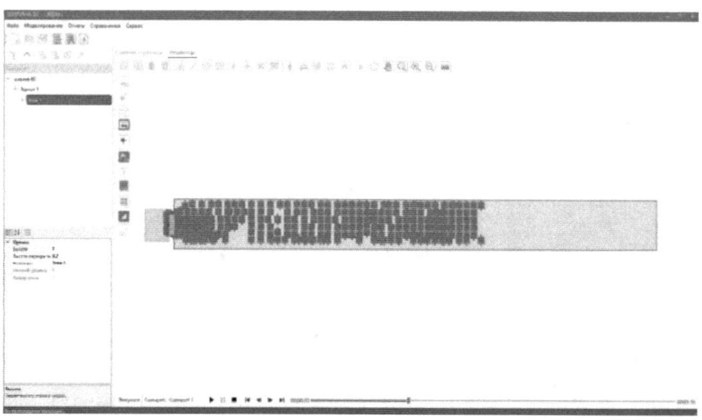

Figure 3. Urban: visualization of the simulation of the human flow traffic through apertures of various widths.

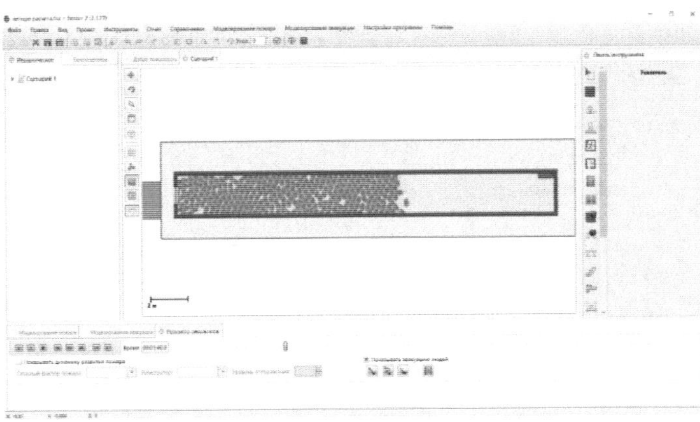

Figure 4. Fenix+: visualization of the simulation of the human flow traffic through apertures of various widths.

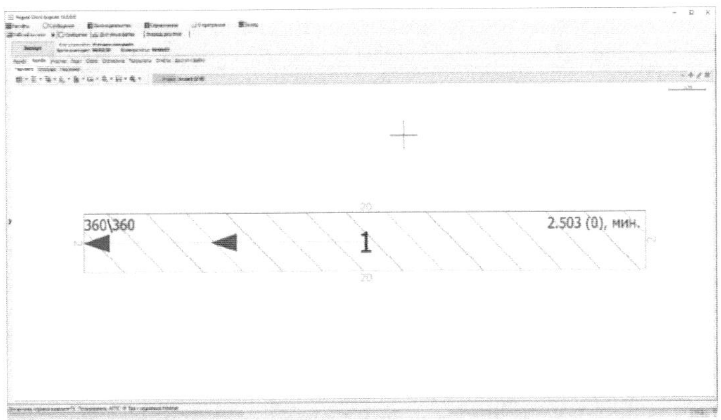

Figure 5. Fogard Pb: visualization of the simulation of the human flow traffic through apertures of various widths.

3. Results and discussions

Based on the simulation results, a comparative table 1 was compiled, which indicates the time of evacuation of people during modeling in various software and computing complexes.

Table 1. The time of evacuation of people from the units.

No.			The width of the door aperture 0.8 m	The width of the door aperture 1,2 m	The width of the door aperture 1,6 m	Horizontal way 2 m wide
1	Manual calculating	-in seconds	490,90 sec	257,14 sec	158,82 sec	79,99 sec
		-in minutes	8,18 min	4,29 min	2,65 min	1,33 min
2	Pathfinder	-in seconds	503,77 sec	258,50 sec	159,50 sec	80,50 sec
		-in minutes	8,39 min	4,31 min	2,66 min	1,34 min
3	Sigma PB	-in seconds	114,25 sec	68,50 sec	58,75 sec	57,25 sec
		-in minutes	1,90 min	1,14 min	0,98 min	0,95 min
4	Urban	-in seconds	76,00 sec	74,00 sec	76,00 sec	63,00 sec
		-in minutes	1,26 min	1,23 min	1,26 min	1,05 min
5	Fenix+	-in seconds	467,00 sec	243,00 sec	148,60 sec	45,40 sec
		-in minutes	7,78 min	4,05 min	2,47 min	0,75 min
6	Fogard Pb	-in seconds	150,20 sec	99,90 sec	74,30 sec	59,50 sec
		-in minutes	2,50 min	1,66 min	1,24 min	0,99 min

Manual calculating is a reference, as it was calculated in accordance with the Methodology. The main drawback of many programs is the deviation of the evacuation time from the reference value, so, we can conclude that the programs do not take into account the patterns of the flow traffic of people. When considering a horizontal section without a door, the Fenix + software complex showed the greatest deviation from the standard (43,24%), while the Sigma PB program showed the largest deviation with a door aperture width of 1,6 m and 1,2 m (63,01% and 73,36%, respectively), with a door aperture width of 0,8 m - Urban (84,52%). A deviation that does not exceed 5% is acceptable. Only one software-computing complex complies with this value (Pathfinder). The remaining programs give serious deviations, which are presented in table 2, and their use for calculations is undesirable.

Table 2. Deviations of the evacuation of people time is calculated in the programs.

No.	Software and computing systems	Deviations of the evacuation time from the time calculated in accordance with the Methodology
1	Pathfinder	0,42-2,62%
2	Sigma PB	28,43-76,73%
3	Urban	21,24-84,52%
4	Fenix+	4,87-43,24%
5	Fogard Pb	25,61-69,40%

When modeling the human flow traffic, certain drawbacks of software and computing systems were identified. In the Pathfinder program, it is necessary to manually set the initial data on the patterns of human flows traffic with an aperture width of less than 1,6 m. Otherwise, the program will consider the "false aperture effect", that is, reducing the actual width of the aperture for a short time due to the fact that people walking from the sides of the horizontal path are repelled from the corners to the center of the door aperture when passing through it, as people tend to avoid being pressed to the corners of the door aperture. In the Sigma PB software and computing complex, the laborious process of preparing the building and creating an evacuation scenario and an inconvenient interface were noted. When placing people in Sigma PB, the program does not allow you to set the necessary density of the human flow, it is also not possible to set the initial data on the laws of people's traffic. Urban combines the drawbacks of the Sigma PB program, and also with the automatic placing of people according to a given number in a given area, the program places people not only in the unit, but also "inside" the envelope building. The qualitative picture of the people's traffic is displayed inadequately, the program does not have the concepts of congestion, merging, decompaction, etc., people move at the same speed, rolling over the heads of people in front, the flow of people merges into a single whole in the door aperture. The Fenix + program does not allow you to set the parameters of the people's traffic with a door aperture width of less than 1,6 m. Fogard Pb does not clearly show the location of doors and people, only their coordinates are in a table form, there is no visualization of the calculation, therefore, it is impossible to fully evaluate the process of human flow's traffic.

Also there are such unsolved problems as the lack of generally accepted criteria for the acceptable quality of programs for evacuation modeling, the lack of qualifications for program developers, the lack of development of the fire models market, since all new developments are cloning of the previous ones, they do not add new approaches, models, or initial data.

4. Conclusions

The results of modeling the human flows trafic are analyzed, deviations are obtained that prove the incorrectness of the algorithms embedded in the software and computing systems designed to calculate the time of evacuation of people. The data obtained give the prospect of finalizing existing programs and show the need to confirm the qualifications of developers of software and computing systems. It is expected that if existing programs are brought into compliance with the Methodology, if leading scientific state organizations publish methodological manuals for developers and designers, if sets of test cases are developed to confirm the qualifications of developers of software and computing systems, it will be possible to enter the public market programs that give adequate results in accordance with the Methodology, taking into account all the subtleties of the dependencies of human traffic.

References

[1] Federal Law of the Russian Federation of 22.07.2008 No.123-FZ "Technical regulation on fire safety requirements" (in Russian)
[2] V. Kholshchevnikov, and A. Parfyonenko, "Comparison of various models of the human flows traffic and the results of software and computing systems," *Fire and explosion prevention,* pp. 68-75, 2015.
[3] Z. Aibuev, I. Isaevich, and M. Medyinik, "Free traffic of people in a flow and problems of individual-flow modeling," *Fire and explosion prevention,* pp. 66-73, 2015.
[4] V. Kolodkin, and M. Galiullin, "Software algorithms that implement the model of the human flows traffic in the control system for the evacuation of people from the building," *Fire and explosion prevention,* pp. 75-78, 2016.
[5] A. Parfyonenko, "Methodology for modeling human flows and the practice of programming their traffic during evacuation," *Fire and explosion prevention,* pp. 46-55, 2014.
[6] V. Kholshchevnikov, "Once again on the modeling of human flows in assessing fire risk," *Fire and explosion prevention,* pp. 61-69, 2012.
[7] The methodology for determining the calculated values of fire risk in buildings, structures and structures of various classes of functional fire hazard. Appendix to the Order of the Ministry of Emergencies of Russia dated June 30, 2009 No. 382, subject to the changes approved by Order No. 632 of the Ministry of Emergencies of the Russian Federation of December 2, 2015 (in Russian), 2009.
[8] A. Dandenko, V. Tarakanov, and V. Shtern, "Application of an individual-flow model of evacuation when assessing fire hazard risks in public buildings," *Reports of Tomsk State University of Control Systems and Radio Electronics, 2014.*
[9] O. Barahovskaya, and A. Perevalov, "On the calculation of the actual time of evacuation of people from the units," *Fire and explosion prevention* , pp. 41-47, 2010.
[10] V. Predtechenskiy, *Human Streams in Buildings,* Moscow: Stroyizdat (in Russian), 1983.
[11] V. Kholshchevnikov, "Models for calculating the evacuation of people become obsolete," *Fire and explosion prevention,* pp. 39-55, 2017.
[12] V. Kholshchevnikov, A. Giletich, D. Ushakov, and A. Parfyonenko, "General pattern of change in the parameters of the human flows traffic of different functional contingent in buildings and structures," *Fire and explosion prevention,* pp. 32-41, 2011.
[13] I. Bakirov, and M. Faizylina, "An individual-flow model of evacuation as an effective method of a probabilistic approach in assessing the risks of fire hazard in buildings," *Problems of ensuring safety during emergency response,* Ufa, Russia, *2016.*
[14] E. Urtaev, A. Vitovtov, and D. Ruskih, "Features of the domestic methodology for calculating the evacuation of people by the individual-flow method, implemented using foreign programs," *Scientific and analytical journal Bulletin of the St. Petersburg University of the State Fire Service EMERCOM of Russia,* Saint Petersburg, Russia, 2018.

The State of Public Space in Small Towns in Terms of its Users´ Risk Groups

Zuzana Kramarova [1], **Ales Kankovsky** [1]

[1] Institute of Technology and Business in České Budějovice, Department of Civil Engineering, 370 01 České Budějovice, Czech Republic

kramarova@mail.vstecb.cz

Abstract. At the beginning of the 3rd millennium, the issue of small towns is still a very problematic area in terms of urban planning which lies on the fringes of the interest of the professional public. The issue of the quality of public space in small towns is not tackled and its current development is in some places unprofessional because it is determined mainly by the local community instead of experts. Public space in small towns is massively subordinated to the needs of individual car transport and as a result, this space is aesthetically and functionally degrading, and at the same time, it contributes to the deterioration of the health of a population that does not move enough especially the most commonly used part - walking. The aim of this paper is to approach the needs of adaptation of public space in small towns for risk groups of the population, which are mainly seniors, children, and people accompanying children under 5 years of age.

1. Introduction - definition of a small town

In the Czech Republic, the structure of human settlement, which includes more than 6,200 municipalities, is relatively uniform worldwide. Division of the Czech Republic into regional areas is shown in Figure 1. Since 2003 these municipalities have been territorially and administratively divided into 14 regions and a total of 205 AD MEP (municipalities with extended powers, the so-called municipalities of the 3rd degree), based on Acts No. 129/2000 Coll., No. 314/2002 Coll. and No. 128/2000 Coll. Municipalities with extended powers, in the vast majority of cases smaller or small towns, are evenly distributed throughout the country so that their accessibility is no longer than about 15 km for the citizens who live there. City or state administration is concentrated in the individual ORPs, such as building authorities, spatial planning authorities, branches of financial authorities, nature protection authorities, population registers and others. These centres are therefore naturally visited by a larger number of citizens.

From the point of view of the legislative division of settlements, according to Act No. 128/2000 Coll., On Municipalities (Municipal Act), four categories are known: municipalities, towns, cities and statutory cities. Their definitions are given by law as follows:

In §1 „*The municipality is a basic territorial self-governing community of citizens; forms a territorial unit, which is defined by the boundary of the municipality.*“, in §3 „*A municipality with at least 3,000 inhabitants is a city if the chairman of the Chamber of Deputies so decides on the proposal of the municipality after the statement of the government or if it was before 17 May 1954."* and *„A municipality is a township, if so determined by the chairman of the Chamber of Deputies at the suggestion of the village, after a statement by the government or if it was authorized before 17 May 1954 as a township."* (sources are abbreviated, not literal). Statutory cities are defined separately in this law in §4: *„Statutory cities are Kladno, České Budějovice, Plzeň, Karlovy Vary, Ústí nad Labem, Liberec, Jablonec nad Nisou, Hradec Králové, Pardubice, Jihlava, Brno, Zlín, Olomouc, Přerov, Chomutov, Děčín, Frýdek-Místek, Ostrava, Opava, Havířov, Most, Teplice, Karviná, Mladá Boleslav, Prostějov a Třinec. The territory of statutory cities can be divided into city districts or city parts with their self-government bodies"* [1].

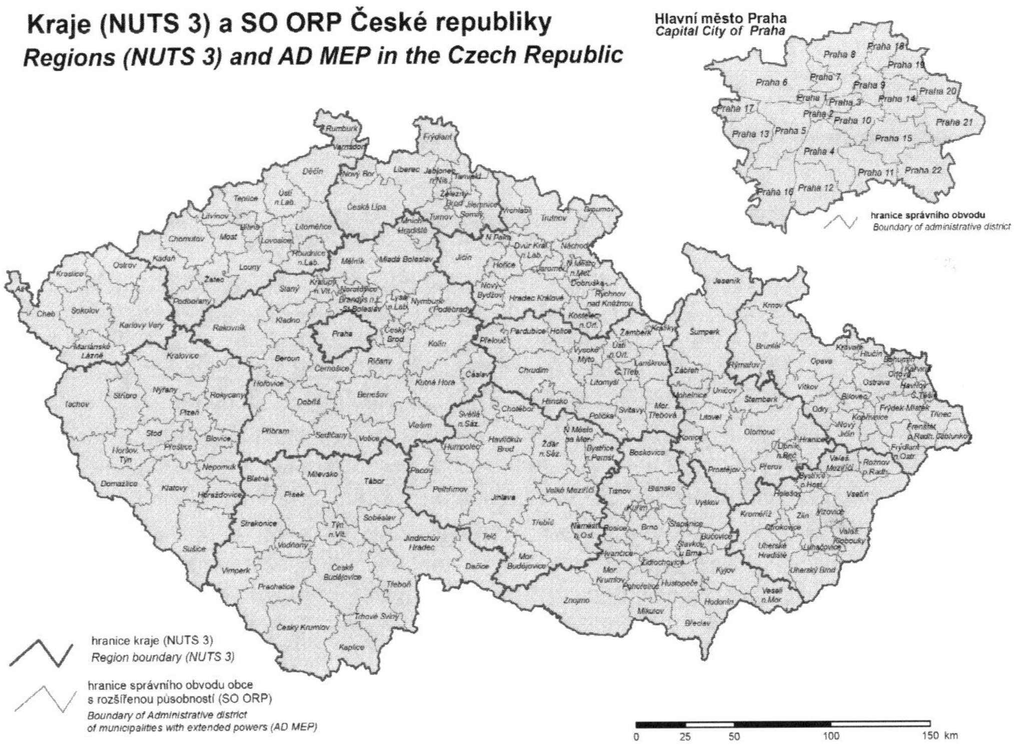

Figure 1. Regions and AD MEP in the Czech Republic.

Another of the primarily monitored criteria is the size of settlements. However, it is not their area, but especially the number of inhabitants who live/have permanent residence here. This information is available in the most up-to-date form online on the website of the Czech Statistical Office [2]. Regarding the classification and classification of individual settlements according to the size/number of inhabitants, according to the CZSO, the division into "size groups of municipalities". According to the CZSO, there are 10 groups of these, which is shown in Table 1.

Table 1. Number of municipalities in size groups – territorial comparison (31.12.2018) [2].

	Total	less than 200	200 - 499	500 - 999	1 000 - 1 999	2 000 - 4 999	5 000 - 9 999	10 000 - 19 999	20 000 - 49 999	50 000 - 99 999	100 000 or more
Czech Republic	6 258	1 423	1 997	1 366	769	426	147	69	43	12	6
Praha, Capital City	1	-	-	-	-	-	-	-	-	-	1
Středočeský Reg.	1 144	227	370	279	152	75	21	15	4	1	-
Jihočeský Reg.	624	229	206	87	49	32	14	2	4	1	-
Plzeňský Reg.	501	163	166	79	49	27	11	4	1	-	1
Karlovarský Reg.	134	16	45	31	15	16	4	4	3	-	-
Ústecký Reg.	354	43	127	89	44	26	8	10	5	2	-
Liberecký Reg.	215	29	64	60	30	17	10	2	2	-	1
Královéhradecký Reg.	448	99	171	98	36	22	14	6	1	1	-
Pardubický Reg.	451	106	172	89	48	19	9	6	1	1	-
Reg.Vysočina	704	324	216	96	37	14	9	4	3	1	-
Jihomoravský Reg.	673	108	193	178	108	61	15	4	5	-	1
Olomoucký Reg.	402	47	127	109	74	32	4	5	3	-	1
Zlínský Reg.	307	19	83	97	53	37	9	4	4	1	-
Moravskoslezský Reg.	300	13	57	74	74	48	19	3	7	4	1

For the purposes of this contribution, it is possible to define settlements with a size of 5,000 - 9,999 inhabitants as small towns. These are cities that are a natural center of a lower level for their surroundings or are directly categorized as AD MEP. Another common very important denominator is the absence of public transport, which thus evokes a higher load of individual transport - especially automobiles. This means cities where pedestrian mobility is significantly more problematic, particularly at-risk groups such as the elderly, children, and adults accompanying a young child, to greater distances, which in larger cities with a functioning public transport system are covered, inter alia, by this type of transport.

These risk groups of the population create specific demands on public space in normal operation - its division, clarity, the addition of furniture, etc.

2. Public space and public area.

In terms of legislation and urban theory, there are two very related terms – public space and public area. These terms are sometimes placed on the level of synonyms, but some professional publications work with them differently.

Act No. 128/2000 Coll., On municipalities, defines only the concept of public area. It is precisely described in §34: *„Public areas are all squares, streets, markets, sidewalks, public greenery, parks and other spaces accessible to everyone without restrictions, thus serving general use, regardless of the ownership of the space."* [1]. The term public space is not defined in this Act or any other, which concerns areas related to spatial planning.

Another legislative basis that operates with the term public areas is Decree No. 501/2006 Coll [3], On general requirements for the use of land. Specifically in §7, which is entitled "Areas of public areas", the requirements for these areas are formulated. Paragraph 1 talks about the reasons why areas need to be demarcated and paragraph 2 specify these areas in more detail: *„Areas of public spaces usually include existing and proposed plots of land of individual types of public spaces and other plots of land related to transport and technical infrastructure and civic amenities, compatible with the purpose of public spaces. For every two hectares of buildable area of housing, recreation, civic amenities, or mixed* *living space, an area of public space related to this buildable area with an area of at least 1000 m² is defined; this acreage does not include roads"* [3].

The concept of the public area is also mentioned in Act No. 183/2006 Coll. on Spatial Planning and Building Regulations (Building Act). Here in §2, which defines the terms used in the Building Act, the term public infrastructure is defined in paragraph 1, letter k). *„For the purposes of this Act, public infrastructure means land, buildings, facilities, namely 1. Transport infrastructure, such as the construction of roads, railways, waterways, airports, and related facilities; 2. technical infrastructure, which is lines and constructions and operationally related technical equipment, such as water mains, reservoirs, sewers, wastewater treatment plants, constructions to reduce the risk of natural or other disasters, constructions and facilities for waste management, transformer stations, power lines, communication lines of the public communication network and electronic communication equipment of the public communication network, product pipelines and gas storage facilities; 3. civic amenities, which are buildings, facilities and land used for example for education and upbringing, social services and family care, health services, culture, public administration, protection of the population; 4.public area established or used in the public interest"* [5].

When defining the term public area, the Building Act and Decree No. 501/2006 Coll. refer to the above-mentioned definition given in Act No. 128/2000 Coll.

However, both terms are defined in the professional texts of the Public Space Office and dealt it with differently: *Public space is considered as a publicly accessible complex physical part of the environment, which in addition to public space are also, for example, accessible publicly space of the city exterior, courtyards and interiors of buildings and public space are understood as a physical part of the environment, which is publicly accessible space floor of the city exterior*. The public area is thus understood here as a subset of public space, which in addition to the area of space also includes the space above it [6]. When defining the term public area, the Building Act and Decree No. 501/2006 Coll. refer to the above-mentioned definition given in Act No. 128/2000 Coll.

Another professional text that deals with the issue are the Methodology of Designing Public Spaces by the authors' Barbara Krátká Adámková and Anna Magni. This methodology was created within the NAKI project with the identification of DF11P01OVV019 - Methods and tools of landscape architecture for land development, which was developed in 2011-2015 by the Faculty of Horticulture in Lednice, Mendel University in Brno and the Faculty of Law, Palacky University in Olomouc. The text defines both mentioned terms and evaluates them as overlays. After consideration, it uses the term public space as a more complex term. It takes definition from Vladimíra Šilhánková from the book Public Spaces in the Spatial Planning Process from 2003: *„Public space can be defined, for example, as all undeveloped spaces in the city that are freely (free of charge) accessible to all residents and visitors to the city, either continuously or with a time limit (e.g. parks closed at night). The basic characteristic of public space is its habitability associated with its use for residents, ie it must serve the inhabitants of the city to run a variety of physical activities (walking, cycling) and residential (sitting, games).“* and expands it with the observations of Markéta Čablová and her team from the publication Quality Public Spaces Methodology of the Creation and Restoration of Public Spaces from 2015 that public space can be understood as the equivalent of English public space [7].

Vladimír Kuta does not work at all with the term public space in the script Theory of Urban Construction from 1996-97. It defines only urbanized space, urban roads, static and dynamic elements forming the organization and composition of cities and on their basis develops the theory of their creation [8].

Also, one of the world's leading contemporary architects and urban planners, Jan Gehl, does not operate at all with the terms public area and public space. It replaces them with the term *urban space*, which equates with the term *space for life* [9]. This simplified presentation makes it easier for the general public to understand the concept of urban space but is harder to grasp for professional use. Firstly, because urban space cannot be applied to a village (here, by analogy, it must be a rural space), and also probably because the publications in which the term occurs are focused on large metropolises/cities, which Gehl deals within his publishing practice [9-11]. This is irrelevant for practical use in the Czech Republic, because there is only one city that can be described as a city from the global point of view - Prague, although even here a lengthy discussion of the parties included in the city or town could be developed. In other professional texts, Gehl more often relies on the terms open space or urban space, which can be translated as open space/free space and urbanized space/public space, however, their exact definitions are missing.

Even the world's leading architect and urban planner Kevin Lynch do not use the terms public space and public area in his book The Image of the City, which dates back to 1960. Instead, it introduces five basic elements, which are paths, borders, areas, nodes, and important elements, with which it defines the general image of the environment of the seat and its possible modification [12].

Based on a survey of the above-cited professional texts, both terms can be more freely translated into English as public areas = urban area a public space = urban space/open space. As space-type connections are more common in the English-language professional literature in general, it is obvious to assume that foreign experts work more often with a term that includes the whole environment, including surfaces and the space above it, as physical parts. At the same time, the current trend can be stated, which seeks to add to the physical parts mentally and subjectively perceived components, which can be collectively referred to as *genius loci* and which are a very important and integral part for users/residents.

From the previous analysis, it is clear that all definitions are more or less similar, but not identical. Therefore, it is necessary to clearly define what concepts and to what extent further work will be done. From the point of view of user comfort, it is clearly desirable that the problem be solved as widely as possible, that is, from as many aspects as possible, which they perceive as a whole. Therefore, it is more appropriate to continue to operate with the notion of public space than with public areas. From the point of view of legislative aspects, it is appropriate to formulate the definition of the term public area as follows:

The public area *is a publicly accessible area in the exterior of the city formed mainly by a physical free surface in the exterior of the city. These are areas of sidewalks, streets, squares, embankments, markets, parks, and other public access areas in settlements.*

Public space should then form a concept superior to public space and should, in addition to physical areas, also include intangible aspects of perception that influence the users of space in the evaluation. Its definition should then be:

The Public space *is a space accessible to the public in the exterior of the city, which consists of the area of public space and free space above it, as physical components and also of intangible components forming the genius loci. Its basic characteristic is its habitability associated with usability, i.e. it must serve residents and users to perform various physical activities (mobility from point to point) and residential activities (life, sitting or standing, verbal and non-verbal communication, social contact, ...). These are areas of streets, squares, embankments, markets, parks, and other publicly accessible areas in the headquarters.*

IOP Conf. Series: Materials Science and Engineering 1066 (2021) 012015 doi:10.1088/1757-899X/1066/1/012015

Both definitions deliberately do not mention property rights in the area, because the public space can be a privately owned area, which is accessible to the public for a certain time - for example, the courtyard of the library or the adjacent garden of a private castle.

3. The needs of risk groups in public space

There are many users of public space: pedestrians, car transport, public transport systems, fauna, and flora. However, in the last 30 years or so, public space has been adapted mainly to car traffic and its enormous increase. Primarily historically, however, it was not traffic, but pedestrians who formed the majority group of users of urban public space. The trend of adapting to traffic has been criticized in recent years by the world's leading urban planners such as Gehl and Lynch [9-12]. The ever-increasing demands of individual car traffic on the width of roads and especially parking areas are alarming and the urban area of cities is gradually becoming a confusing and dangerous zone for other groups of users.

The primary creator and user of public space is a man. Its needs and requirements are key to the creation and modification of public space. Here, however, it is necessary to specify the person in more detail, because individual groups of people often have not only very different but sometimes completely conflicting requirements.

We can divide people into individual groups from several points of view. For example, Čablová and the team state the division into these groups: pre-school children, primary school children, teenagers, and young people, adults and workers, men and women, parents or guardians with young children, the unemployed, locals, pensioners and the elderly, people with disabilities, migrants and other minority ethnic groups, tourists and visitors [13]. This is a mixed division according to the age of individuals, their job classification, lifestyle, and other aspects. On the other hand, Krátká Adámková and Magni do not solve the problem of dividing users at all [7].

The division of people according to age categories, abilities, and speed of their movement is one of the ways of categorization based on gross age groups and possibilities of their walking movement, respectively on their different requirements for quality, dimensional and spatial arrangement of public space. At present, it is necessary to focus mainly on risk groups of the population that are not treated legislatively and normatively. Such risk groups include, in particular, persons with reduced mobility or orientation. According to Decree No. 398/2009 Coll. these are people with physical, visual, hearing, and mental disabilities, elderly people, pregnant women, people accompanying a child in a pram, or a child under three years of age [3, 14].

Persons with physical disabilities require very specific modifications of the ground floor, which the above-mentioned decree clearly defines - tactile strips, signals, sound signals, lowering the height levels, increasing the space for the possibility of turning a wheelchair, etc. However, other groups are also partially neglected in this decree, for example, the size of the set "person with a child in a stroller" is completely different from the size of a person in a wheelchair, yet the decree does not specify specific requirements. Also when determining viewing areas for a safe crossing, it should be taken into account that the person pushing the pram in front of him is much further away from the edge of the road and therefore the top of the viewing area should be placed at a different point than standard.

For the purposes of mapping, the overview of the possibilities or limits of some routes in the area, the so-called generals of barrier-free routes of the city are processed in the cities. These are spatial planning documents at the level of spatial study according to Act No. 183/2006 Coll. (building law) [5]. The Building Act and its implementing decree No. 500/2006 Coll. however, they do not specify the exact requirements or the content of this document, therefore the contents may differ according to the assignment and the processor.

The mobility system is currently dealt with by both the Ministry of Transport of the Czech Republic and the Ministry of Regional Development of the Czech Republic. In December 2015, the Ministry of Transport prepared and issued a document entitled Methodology for the Preparation of Sustainable Mobility Plans for Cities in the Czech Republic, where public space is addressed under Activity B.4 - Public Space. It should be noted, however, that the whole document sets out the methodology for drawing up transport plans as a whole. Pedestrian, bicycle, and public spaces are therefore only one of many aspects in this document, and none of them are among the priorities. Overall, the document focuses on the effort to balance the various components of transport and the main focus is again on motorized road transport.

It is also interesting to watch the above-mentioned groups in terms of the frequency of pedestrian movement in public space. The groups that most often move on foot in the area (as determined by a local survey) are children, teenagers, seniors, pedestrians with prams, and pedestrians with children (ranked from the smallest group to the less numerous). Other groups are very under-represented on foot. Although some groups are more massively represented in the territory in terms of frequency and number than others, space is not adapted to the children, seniors. To a limited extent, the safety of children is taken care of by completely different forms of regulation, such as the presence of an authorized person to supervise crossing at critical times at very busy crossings and streets. However, these measures have a short-term, limited effect and do not address safety as a whole and conceptually.

The adaptation of public space for seniors is also very neglected. These are mobile people but over shorter distances. A senior with increasing age needs places on his routes where he can rest and then be able to continue the route. The most suitable element of urban furniture for this purpose is a bench. However, equipping the public space of small towns with benches and other furniture is very low. Benches or seats can be found in parks in the area of central squares, possibly at bus stops. But their placement on the routes of seniors that they routinely perform - trips to shopping, to the doctor, or to the office is almost zero. In many small towns, this furniture is not funded, there is a lack of space for its location and corresponding studies locating its location.

Another element that is necessary for the actual implementation of walking in a public space is a safe walking zone. It is mainly represented by sidewalks or reserved lanes along roads. The minimum width of the pavement is given in ČSN 73 6110 - Design of local roads in chapter 10.1. This width is given by a value of 1.5 meters but does not take into account numerically / coefficient the intensity of the pedestrian current. This shortcoming is widely used in the adjustment of street widths in small towns, where they often expand the passage profiles of lanes or the implementation of parking spaces at the expense of the width of sidewalks, such as Chotěboř in Krále Jana Street and T. G. Masaryk Square [15].

Another relatively common phenomenon is the disruption of pedestrian crossings when making modifications to public spaces in small towns. This is mainly due to an increase in traffic intensity on roads and a change in view conditions. Strict care is taken to observe the observation triangles, but their size has increased over the last 30 years. This leads to the fact that the existing transition does not comply with current legislation and its relocation or direct cancellation is required during construction modifications.

4. Conclusion

The authors of the paper surveyed the current state of the legislative and normative framework of the issue of population mobility in the Czech Republic. They also focused on a survey of the current state of public space in small towns, which was defined by a population in the range of 5,000 - 10,000 inhabitants. A walking survey of the central parts of these cities was carried out in these cities, followed by an evaluation of the quality of the public space, supplemented by a list of the most common shortcomings and problems of this space. Among the most frequently monitored, it is necessary to mention the insufficient width of pedestrian roads in the context of user intensity, a small number of absolute absence of urban furniture (especially benches), which will improve public space and facilitate its use for seniors, the insufficiently designed intersection of pedestrian transport in terms of safety of risk groups, etc.

It was clear from the above that in the Czech Republic, public space in small towns is relatively neglected in terms of quality. This situation can be evidenced, among other things, by the decline in the number of pedestrians from the ranks of the productive population, which can most easily choose between several available modes of transport.

Furthermore, it was found that there is no uniform basis or methodology that would conceptually address the issue of pedestrian movement in public space for the entire country. This is left to the discretion of the municipalities in the individual cities, and where there is a lack of funding, this issue is very neglected.

There is also no normative regulation that would responsibly set minimum requirements for the width and equipment of pedestrian roads in the context of the typological classification of public space (squares, main streets, embankments, side streets, parks, etc.) and responsibly take into account the intensity of the number of users over 24 hours. or at critical peaks during the day.

These facts need to be rectified as a matter of urgency so that municipalities and responsible officials can be given a suitable tool and guidance on how to improve the current unhappy situation and thus make public space not only pleasing but high quality and safe.

References

[1] Czech. Act No. 128/2000 Coll, of May 12th, 2000, on Municipalities (Municipal Act), 2000.
[2] Czech Statistical Office, Accesible from: https://vdb.czso.cz/
[3] Czech. Decree No. 501/2006 Coll, of November 10th, 2006, on general use requirements, 2006.
[4] Czech. Decree No. 268/2009 Coll., of August 12th 2009, on the technical requirements for buildings, 2009.
[5] Czech. Act No. 183/2006 Coll, of March 14th, 2006, on town and country planning and building code (Building Act), 2006.
[6] Public Space Office. *Prague Public Space Design Manual*, Prague, Prague Institute of Planning and Development (In Czech), 2014.
[7] B. Krátká Adámková, and A. Magni, *Designing public spaces in newly set up residential complexes - with the emphasis on the garden and landscape architecture* (In Czech), 2015.
[8] V. Kuta, *Teorie stavby měst a urbanismus*. Ostrava, VŠB – TU Ostrava (In Czech) 1996.
[9] J. Gehl, and L. Gemzoe, *New City Spaces*, 2002.
[10] J. Gehl, *Life Between Buildings: Using Public Space*, 2000.
[11] J. Gehl, *Cities for People*, 2012.
[12] K. Lynch, *The image of the city*, 2004.
[13] M. Čáblová, M. Maceková, L. Mlčák, M. Nawrath, M. Římanová, R. Sedlák, and P. Šilberská, *Quality of Public Space: Methodology of creation and renewal of public spaces* (In Czech), 2011.
[14] Czech. Decree No. 398/2009 Coll., of November 5th 2009, on general technical requirements to secure barrier-free usage of buildings, 2009.
[15] Czech. *CSN 73 6110 Design of urban roads* (in Czech), 2006.

IOP Conf. Series: Materials Science and Engineering 1066 (2021) 012016 doi:10.1088/1757-899X/1066/1/012016

Thermal Properties and Overheating of Ventilated Air Cavities Under Pitched Roof Coverings

Ondřej Pilný[1], Lubor Kalousek[1]

[1] Institute of Building Structures, Brno University of Technology, Faculty of Civil Engineering, Veveří 331/95, 602 00 Brno, Czech Republic

pilny.o@fce.vutbr.cz, kalousek.l@fce.vutbr.cz

Abstract. Ventilated air cavities have been a mandatory part of thermally insulated pitched roofs in the Czech Republic. The purpose of these important elements is, among other things, primarily to prevent condensation of water vapor, which could otherwise endanger the roof structure and thus negatively affect the life of the entire composition. However, their thermal properties are often neglected and their overall design is overwhelmingly approached, due to the very difficult and complex calculation, as an empirical and normative matter without a higher link to the actual properties of the air cavity itself. It is therefore necessary that mentioned properties must be investigated more thoroughly, especially with increasing demands on thermal properties and increasingly extreme temperature fluctuations. This work describes a real full-scale experiment in which the properties of temperature and humidity of ventilated air cavities at the boundary with the external environment were measured. The composition of the experimental roof of the building was measured in all important layers, together with all the external and internal environmental conditions. The output is then a detailed course of behavior, which was analyzed and described.

1. Introduction

Roofs are, from a constructional and thermal-technical point of view, one of the most important parts of buildings. From a historical point of view, we encounter sloping or steep roofs. At present and in the conditions of the Czech Republic, it is always possible to find a very similar roof composition, especially of the upper structure, which consists of the so-called additional waterproofing layer (hereinafter AWL), battens and roofing. Thanks to battens, a ventilated air cavity is created. This space serves for the drainage of infiltrated water and for the prevention of condensation of water vapor. Thus, it can affect the durability of the roof itself and can also significantly affect, by its very nature of the i.e. open-air cavity, the thermal-humidity flow of the roof structures. It is therefore a completely critical mechanism responsible for the proper functioning of the roof. As for the design of this cavity, the detailed calculation is very complicated and complex [1]. Therefore, in common practice the problematic itself is approached purely as a norm matter [2], which considers the assumption that the cavity has the same properties as the external environment. However, this approach is often inaccurate for more complex roofs and does not correspond to the real behavior and characteristics of these spaces [3, 4]. It is therefore necessary to examine the properties of these cavities more thoroughly, especially with respect to the growing requirements for thermal insulation properties and increasingly extreme temperature fluctuation [5].

This paper describes a full-scale experiment performed on a real object, where sensors are installed in the relevant critical places of the roof cladding composition, including the critical air cavity, and the thermal-humidity flow and state of this environment is monitored. The roof cladding is oriented sharply to the south and its slope is 30°. The composition and dimensions of the air cavity are in accordance with the Czech Technical Standard [2]. The roof has heavy structure with thermal insulation over rafters. For an accurate description of the experimental structure, the measuring system and its installation, see [6]. At the same time, the boundary conditions of the exterior are monitored.

The paper also focuses on the analysis of the data using appropriately selected software and verification of the hypothesis that this cavity does not behave as an external environment whose parameters are attributed to it and aims to prove that in critical months the cavity can both negatively and positively affect the heat transfer mechanism.

2. Methodology

2.1. Data collection of thermal-humidity of the air cavity and the external environment

For a data collection, a measuring system [6] consisting of thermocouples, humidity sensors and data loggers is installed in the roof of the experimental building. To monitor the parameters of the cavity, the thermal-humidity sensor is installed, see Figure 1. Due to the importance of the external environment parameters, data from the meteorological station CHMI B2POHO01 [5] located near the experimental object are used.

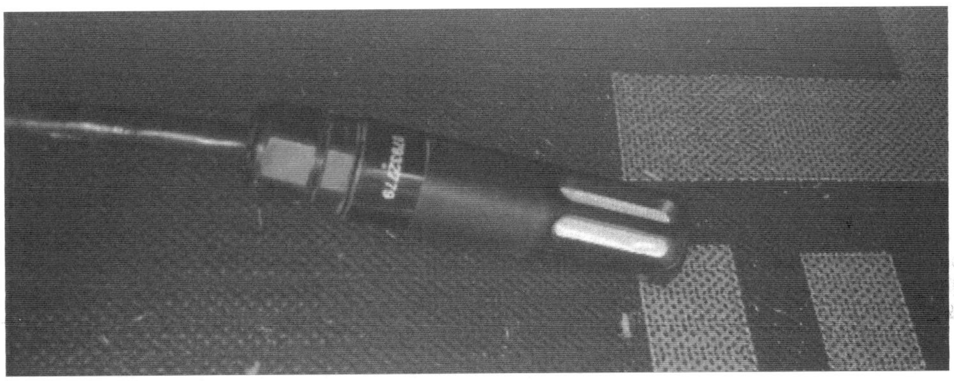

Figure 1. Thermal-humidity sensor installed in air cavity [7].

Figure 2–4. Thermocouples on the AWL and on lower surface and upper surface of the roofing [7].

To determine the complex behavior of the cavity together with the influences of the external environment, thermocouples are installed in critical places, see Figure 2 – 4. The sensor on the upper surface of the roofing is covered with a next layer of roofing to eliminate data distortion by direct sunlight. The sensors are calibrated before installation.

2.2. *Monitored parameters*

The parameters that are monitored and compared by measuring system include the temperature and humidity of the exterior ventilated air cavity, which is further extended by the temperature on and under the roofing and the temperature on AWL. CHMI parameters that are monitored and compared include exterior air temperature, humidity and ground temperature. Data are recorded at 10-minute intervals.

2.3. *Time horizon*

To compare the benefits and negatives of the ventilated air cavity of the roof, the months January and July are selected according to [5], which statistically appear to be the coldest and warmest months when comparing temperatures in the Czech Republic from 2010 to 2019. January has an average temperature of -1.4°C and July 18.9°C. Furthermore, the critical norm date of 21st August is selected, on which the course of parameters is monitored in detail [8].

2.4. *Hypothesis demonstration*

To prove the hypothesis of different parameters of the ventilated air cavity (hereinafter VAC) than the exterior, the program MS Excel, visual analysis and computational analysis are used, which compares the average values of temperature and humidity on the individual days. If the temperature of the VAC in January is higher than the exterior air and ground temperature, temperature of the VAC in July is higher than the exterior air and ground temperature, humidity of the VAC in January lower than the air humidity and the humidity of the VAC in July lower than the air humidity, the hypothesis can be considered as proved. For further confirmation of the hypothesis, the norm day of 21st August is selected and detailed course of all monitored parameters according to point 2.2 is recorded and analyzed.

In the case of a norm day, the complex course of temperatures of all relevant parts of the upper structure of the roof is monitored due to the hypothesis of the effect of a cumulative temperature increase. The effect is associated with the specific heat capacity of the roofing and the radiation of heat back into the air cavity even in the case of cloudy skies or night hours. This phenomenon hypothetically leads to the different behavior of the cavity under investigation.

3. Results

3.1. *Graph of the course of temperatures and humidity in January* – See Figure 5.

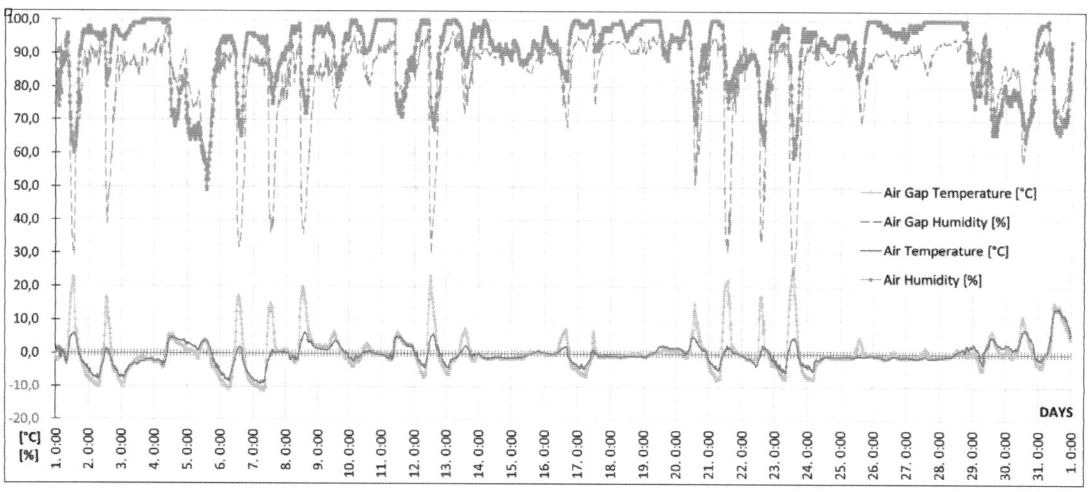

Figure 5. Graph of the course of temperatures and humidity in January [7].

3.2. Graph of the course of temperatures and humidity in July – See Figure 6.

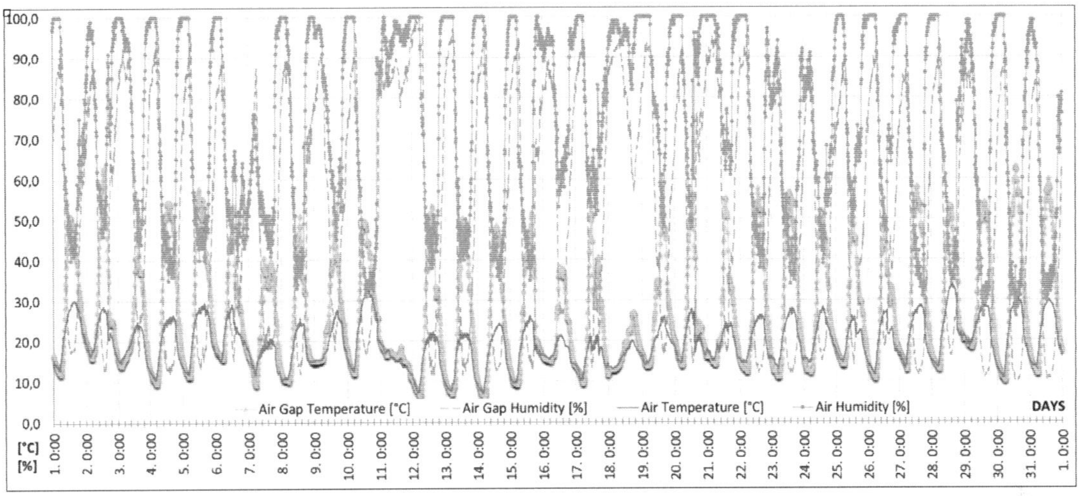

Figure 6. Graph of the course of temperatures and humidity in July [7].

3.3. Graph of the course of temperatures and humidity in critical norm day 21st August – See Figure 7.

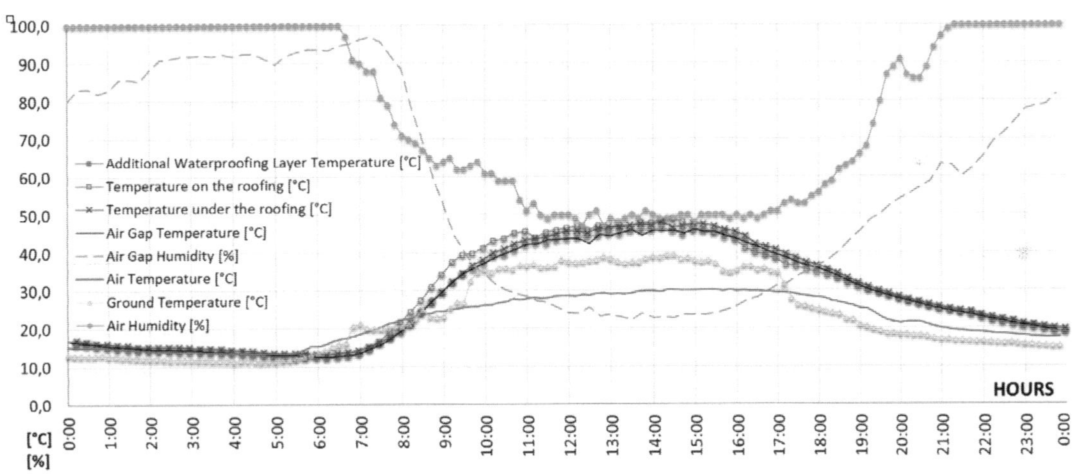

Figure 7. Graph of the course of temperatures and humidity in critical norm day [7].

3.4. Comparison of parameters of the month of January

Table 1. Comparison of Different Parameters of Air Cavity and Air in January.

	Air Cav. Temp.	Air Temp.	Air Cav. Humidity	Air Humidity
Average	0.6076 °C	-0.1000 °C	83.9641 %	90.3988 %
Max	24.3 °C	14 °C	98.3 %	100 %
Min	-11.2 °C	-9.2 °C	23.5 %	49 %
Average Deviation	3.6284 °C	2.2672 °C	8.9592 %	8.01917 %

3.5 Comparison of parameters of the month of July

Table 2. Comparison of Different parameters of Air Cavity and Air in July.

	Air Cav. Temp.	Air Temp.	Air Cav. Humidity	Air Humidity
Average	27.0915 °C	19.3763 °C	54.7988 %	74.1113 %
Max	63.3 °C	33.7 °C	100 %	100 %
Min	6.2 °C	7 °C	9,6 %	26 %
Average Deviation	12.2972 °C	4.5122 °C	26.3877 %	20.6399 %

4. Discussion

4.1. Analysis of the course of temperatures and humidity in January

As can be seen from the graph, see Figure 5, the air cavity temperature repeatedly exceeds the ambient temperature by several times during daily hours in the course of the month. In these cases, the cavity is heated by radiating of the roofing, which in turn obtains solar gains in clear and sunny weather. This effect is partially transferred to the night hours, when it is possible to count on the short-term radiation of the roofing and the associated positive effect of the cavity to reduce heat loss due to a lower temperature difference between the exterior and interior. At night and in the absence of solar gains, however, the effect is the opposite and, on the contrary, leads to cooling of the air cavity and thus increase of heat leakage, i.e. a negative effect. Furthermore, according to the graph see Figure 5, the humidity of the exterior is in most cases, including the transition period of day and night, higher than the humidity of the ventilated air cavity. Table 1 also shows the average temperature difference across the entire observed period, when a higher temperature inside the cavity is evident together with lower humidity. At the same time, it is clear that due to solar gains, this space shows higher deviations both in temperature and humidity.

4.2. Analysis of the course of temperatures and humidity in July

As can be seen from the graph, see Figure 6, the air cavity temperature again exceeds the ambient temperature several times during the daily hours, and partly also nightly hours throughout the entire month. The mechanism of solar gains is the same, but this time with negative effect that leads to overheating of the cavity and thus increase the negative heat gain from exterior. This effect is transferred into air cavity even into the late-night hours, see Figure 7, which prevents the cavity from cooling down and leads to a cumulative effect in the long run. According to the graph, the humidity of the exterior is again in most cases, including the transition period of day and night, higher than the humidity of the ventilated air cavity, which can be attributed to the much higher enthalpy of air in the cavity. Table 2 shows the maximum extreme temperature of the cavity, which exceeds the exterior parameters of the air by almost 30 °C. This temperature leads to overheating of the entire roof cladding. At the same time, a higher temperature and humidity deviation is again apparent, which is again caused by solar gains.

4.3. Analysis of the course of temperatures and humidity in critical norm day 21st August.

The graph, see Figure 7, shows the above-mentioned cumulative effect, which is amplified by high solar gains, which are then radiated into the roof ventilated air cavity almost all night. As we can also see, the temperature of the air cavity together with the temperatures of the roofing and the AWL is higher than the air temperature and than the ground temperature surrounding for the whole monitored day. As can be seen from the graph, this change occurs only for a short period of time during the early morning hours, when, in addition to temperature, there is also a change in the course of humidity due to the lower enthalpy of air in the cavity.

5. Conclusion

The following is evident from the discussion chapter:

- The hypothesis regarding the different behavior of the ventilated air cavity of the roof is confirmed according to the previous chapter for both winter and summer period. In both cases, due to solar gains, the roofing heats up, which then radiates heat into the cavity itself, which increases the temperatures.
- The hypothesis regarding the positive effect of the ventilated air cavity on the heat transfer mechanism in the winter months is confirmed. Due to the higher temperature in the cavity during sunny days, a minor temperature difference occurs and thus the heat loss from the interior is lower. This contributes, assuming the sufficient solar gains, to the stability of the interior temperature during the winter months and at low exterior temperatures.
- The hypothesis regarding the negative effect of the ventilated air cavity on the heat transfer mechanism in the summer months is confirmed. Due to the higher temperature in the cavity caused by the radiation of the roofing with often high specific heat capacity, thanks to the material of its production, a cumulative increase in temperature can be observed from the graph of the summer month. This leads to a constant increase in the temperature of the upper structure of the roof, which leads to destabilization of the interior temperature in the summer months and the need to artificially cool the interior.

From the text above, it is clear that the outcome depends on the following factors:

- The roofing material, as well as the color of both upper and lower surfaces, can both positively and negatively affect the behavior of the ventilated air cavity. The high specific heat capacity of the roofing material can retain the heat for a longer time period, which is then radiated into the cavity and thus derive one of the above-mentioned mechanisms.
- The exposure of the roof cladding to weather conditions has a direct effect on the functionality of the above-mentioned mechanisms.
- Ventilation of the air cavity and the upper roof cladding has a direct effect on the functionality of the above-mentioned mechanisms. Theoretically, it can be expected that with a higher dimensioning of the ventilated air layer, or thorough ventilation, the development of temperatures in both mechanisms (cooling/heating) will slow down and thus approach the parameters of the exterior with temperatures and humidity. At the same time, it can be expected that if the roof is of a suitable slope, the resulting stack effect will positively stabilize the entire cavity mechanism [9, 10].

It is therefore clear from the above text that the influence of ventilated air cavities of pitched roofs must be thoroughly investigated. Not only do they not approach the exterior parameters most of the time, the properties of which we attribute to them, but they can both negatively and positively affect the functionality of heat transfer through the roof and thus fundamentally contribute to both stabilization and destabilization of the indoor environment.

For future research in this field, it is necessary to thoroughly analyze the effect of the height of the ventilated air cavities, as well as the slope of the roof in their parameters. At the same time, it is also important to focus, in addition to the months listed here, on a comprehensive analysis of the behavior of these layers throughout the year and thus eliminate accidental weather conditions. Research should also address the issue of cavity behavior in the so-called transitional spring and autumn months, when the fluctuations in ambient temperature are very variable and unstable for a long time.

A completely separate direction for the future research should be the analysis of the influence of the roofing material itself and the tightness of the roofing locks and joints, which can again may have a direct effect on the functionality of the cavity mechanisms.

Acknowledgement
This paper has been worked out under the project No. FAST-J-20-6275 “Analýza hmotných konstrukcí šikmých střech a jejich vliv na stav vnitřního prostředí".

References
[1] H. M. Künzel, A. N. Karagiozis, and M. Kehrer, “Assessing the benefits of cavity ventilation by hygrothermal simulation,” *Proceedings Building Physics Symposium in honour of Prof. Hugo Hens,* Katholieke Universiteit Leuven, Leuven, Belgium, pp. 17–21, 2008.
[2] ČSN 73 1901-1, 1901-2 Designing of roofs – Basic provisions, Roof with discontinously laid roof covering, Praha ÚNMZ, 2020.
[3] N. S. Bunkholt, T. Säwén, M. Stockhaus, T. Kvande, L. Gullbrekken, P. Wahlgren, and J. Lohne, “Experimental Study of Thermal Buoyancy in the Cavity of Ventilated Roofs,” *Buildings,* 2020.
[4] A. Gagliano, F. Patania, F. Nocera, A. Ferlito, and A. Galesi, “Thermal performance of ventilated roofs during summer period,” *Energy and Buildings,* vol. 49, pp. 611-618, 2012.
[5] Portál ČHMÚ : Historická data : Počasí : Územní teploty. Portál ČHMÚ : Home [online]. Avaible online at: http://portal.chmi.cz/historicka-data/pocasi/uzemni-teploty
[6] O. Pilný, “Experimentální sběr dat pro zjištění vlivu hmotných plášťů šikmých střech na vnitřní mikroklima,” JUNIORSTAV 2020. *SBORNÍK PŘÍSPĚVKŮ PROCEEDINGS*, pp. 79-84, 2020.
[7] Author’s Archive.
[8] ČSN EN ISO 52017-1, Energy performance of buildings - Sensible and latent heat loads and internal temperatures - Part 1: Generic calculation procedures, Praha ÚNMZ, 2018.
[9] V. Bianco, A. Diana, O. Manca, and S. Nardini, “Numerical investigation of an inclined rectangular cavity for ventilated roofs applications,” *Thermal Science and Engineering Progress*, vol. 6, pp. 426-435, 2018.
[10] E. Cuce, F. Sher, H. Sadiq, P. M. Cuce, T. Guclu, and A. B. Besir, “Sustainable ventilation strategies in buildings: CFD research,” *Sustainable Energy Technologies and Assessments*, vol. 36, 2019.

IOP Conf. Series: Materials Science and Engineering 1066 (2021) 012017 doi:10.1088/1757-899X/1066/1/012017

Digital Model of an Existing Building a Wild Riverbed in Tokyo

Martin Dedic [1]

[1] Czech Technical University in Prague, Faculty of Civil Engineering, Department of Applied Informatics, 166 29 Praha 6 - Dejvice, Czech Republic

martin.dedic@fsv.cvut.cz

Abstract. This paper presents an application of photogrammetric methods for so-called image based modelling and rendering. Digital photogrammetry technology was used for the real use of a digital model of an artificial riverbed for a wild river in Tokyo. This technology was applied using two software programs, where we evaluate their compatibility and propose measures that should increase efficiency in the future in the creation of digital models of existing buildings. We describe the process of creating a model in both software separately and evaluate their use for a given type of object. In this paper, we describe the individual processes from the selection of photographs through the cleaning and smoothing of the point cloud to the final model. We present the practical use of the created digital model. In the pictures we show the solved structure, the process of joining individual photographs and the final model for the overall idea of the application of this method. The results and discussions section describes the evaluation of both software and proposals for measures for their use, especially for water structures of a linear nature containing recurring elements such as water barriers.

1. Introduction

Many digital model creation processes [1, 2] focus on projects of new buildings. An Example of an apartment building design process can be found in [3]. Examples of road-building construction processes can be found in [4, 5, 6]. A digital model of new building provides detailed information, which is very useful through the whole building lifecycle, for the future existence and operation of the building [7, 8].

Currently, there are various methods for creating a digital model of an existing building. At the Department of Engineering Informatics at the Czech Technical University in Prague, we deal with the creation of digital twins of objects with high accuracy. In the previous paper, we described the methods of 3D scanning from pre-scanning preparation through data processing from a 3D scanner to the final digital model of the building.[9] In this paper we will present the application of the photogrammetric method, specifically ground photogrammetry. Of all the optical methods, terrestrial photogrammetry is the least device-intensive (common digital camera) and is therefore the cheapest optical method for obtaining digital data.

It can be expected that importance of the acquisition of 3D models of existing buildings will be growing with continuous implementation of industry 4.0 principles [10] and a necessity to acquire data in BIM formats such as IFC [11]. BIM models of buildings and their accuracy will help not only to reduce failures rate in the building project, time and financial savings during construction, but especially in the operation of the building through its whole lifecycle [12].

2. Digital model creation process

The aim of our research, set based on priorities and scope [13], was to find easy to use and inexpensive technology for the construction of a wild riverbed in Tokyo Olympic in Figure 1. The reason for creating the digital model was disputes over the placement of obstacles on the race track. Water barriers affect the flow of water and thus the difficulty and safety of the plant itself. Obstacles can be moved by rails and moved by their height using additional parts. During the construction design, a race track was designed, including water obstacles, by experts from the Czech Technical University, but the organizers of the Olympic Games arbitrarily changed some obstacles. It was necessary to compare both variants using digital simulation.

A Canon EOS 250D digital camera with a resolution of 24.1 Mpx and an 18-55 mm lens was used to take photographs. A total of 1800 photos were taken. The images were divided into groups according to individual parts of the wild riverbed into 3 parts (starting part, middle part, finish part). In the target part there was a water surface, which we evaluated as potentially problematic in terms of high surface reflectivity. After dividing and deleting the photos, there were 1200 left for the possibility of creating a digital model.

Figure 1. General view of the construction of an artificial riverbed for a wild river in Tokyo.

At the Department of Engineering Informatics, we had software in two options available for image processing using photogrammetric methods. In the first case, we created a digital model using Autodesk Recap Photo software. This software allows you to upload a maximum of 200 photos at a time. For this reason, the photos were divided from the original three into six groups. We divided each part into the right and left bank of the riverbed. The individual parts of the model were of good quality after processing the photographs, but the part of the target part of the riverbed, where the water was located, was not drawn correctly. Before connecting the individual parts, it was necessary to clean the point clouds from digital noise, using the automatic function and manual cleaning. Furthermore, the individual parts were smoothed to make the connection as simple as possible. Puncture points were not used during the photo shoot and therefore it was not possible to combine the individual generated parts of the digital model into one whole. We had to use manual connection using the points we specified, which were always in both connected parts. The character of the building - an artificial riverbed for racing on a wild river - did not allow manual connection of all parts, because it is a line building with constantly recurring water obstacles. It was not possible to distinguish which obstacle is located in which part of the racing riverbed. We consider this attempt to create a digital twin to be unsuccessful.

The second software we have is Reality Capture, which also allows automatic processing of images into a digital model. For the first attempt, we divided the photos into two groups according to which part of the riverbed was the water surface. A total of 1200 (900 + 300) photographs were used. The first group was without water surface, the second with water surface. The water surface was located in the finish part of the riverbed. After generating both parts of the model, process you can see in Figure 2, we had to clear the data of data noise and combine both parts into one final model. Although both parts were processed in sufficient quality, it was not possible to combine them purely into one whole. In the part of the riverbed where the water level began, there were different lighting conditions during the rental of the pictures. Due to this, it was not possible to combine the model into one whole.

Figure 2. The process of creating a point cloud in Reality Capture software.

The quality of the digital model in the second software was sufficient to simulate the flow of water. We had to change the technique of creating a model so that we would create only one model without having to connect to another part. So we uploaded to software 200 photos from the right bank of the starting and middle part of the race track. After a thorough inspection of the generated model, we gradually uploaded more images of the right bank of the finish part, which contained the water surface. This technique didn´t work, so we took photos with a water surface again and chose photos from the left bank of the starting and middle part of the riverbed. This part of the model was processed correctly on the first attempt. Other photos - those that contained the water surface, we added manually with the exact designation, always 3-4 points identical to the points that contained the already created part of the model as well as the inserted photo. This process you can see in Figure 3. Because the water surface was also the first part of the riverbed, especially in the places of obstacles that retained water, it was necessary to use this technique in these places. This process has worked well. The point cloud was complete. Furthermore, the data had to be manually cleaned of digital noise and then smoothed using the automatic function.

Figure 3. The process of sequentially uploading photos.

3. Results and discussions

The final model in Figure 4 consists of only 800 of the 1800 photos taken. From the experience of other projects focused on the creation of digital twins using digital photogrammetry, we know that appropriate preparation before the actual shooting is the most important aspect of the whole process. It is necessary to precisely plan the process of taking photographs, especially for larger or linear constructions. Furthermore, the setting of the camera angle, focal length and, last but not least, lighting conditions. Outdoor photography is unpredictable, but that is why it is necessary to carefully plan the photography so that the shortest possible time runs so that the lighting conditions do not change much during the photo shoot. In this particular case, it would be possible to use aerial photogrammetry using a drone, which we unfortunately didn´t have at the time of the photo shoot.

There are more software that supports digital photogrammetry, but each works a little differently, so you need to take photos according to the capabilities of the software used. The last thing that would be good to use are puncture points, which would make it easier to connect the model. Especially for objects that have a large number of the same or very similar parts such as flat surfaces or repeating masses (in our example, water barriers).

Figure 4. Finished digital model in Reality Capture software.

4. Conclusions

From the presented results it can be said that both software can be used for the creation of digital models. The accuracy of the model is the same but if the model contains glossy surfaces the Reality Capture software is significantly better. In addition, this software is more suitable for large objects that need more than 200 photos, which is the maximum possible number per upload in Autodesk Recap Photo software. For example, a lower number of photographs is sufficient for a digital model of a sculpture, but for larger objects such as buildings, the number of photographs is usually in the hundreds. Both software are suitable for editing the point cloud and then exporting the data.

The price for digital photogrammetry, as opposed to 3D scanning, is mainly supported by the price, when the price of the equipment needed to take photographs is significantly lower than the purchase price of a 3D scanner. Furthermore, the evaluation of data is largely automatic - certainly in the example we selected, it was necessary to connect part of the model manually, but only because it was a very demanding object.

Digital photogrammetry can be used as well as 3D scanners in construction. In particular, linear and tall buildings will be possible to better and more accurately model using aerial photogrammetry, especially today, when drones are becoming more available. The future is a combination of both of these methods. The digital model created in this way can serve as a basis for a building information model (BIM).

Acknowledgment(s)
This work was supported by the Grant Agency of the Czech Technical University in Prague, grant No. SGS20/103/OHK1/2T/11

References

[1] J. Myslín, and J. Kaiser, "State based milestones in process modeling methods based on

behavioral approach", *Education Excellence and Innovation Management through Vision 2020: From Regional Development Sustainability to Global Economic Growth*, vols. I-IX, pp. 3598-3606, 2017.

[2] J. Kaiser, "Process Modeling for BIM," *Central Europe towards Sustainable Building 2016 - Innovations for Sustainable Future*, pp. 781-788, June 2016.

[3] K. Prušková, and J. Kaiser, "Implementation of BIM Technology into the design process using the scheme of BIM Execution Plan," *IOP Conference Series: Materials Science and Engineering*, vol. 471, 2019.

[4] J. Rádl, and J. Kaiser, "Realization Processes of Road-building Projects in the Czech Republic: Necessary Information to Execute Processes," *IOP Conference Series: Materials Science and Engineering*, vol. 603, 2019

[5] J. Rádl, and J. Kaiser, "Benefits of Implementation of Common Data Environment (CDE) into Construction Projects," *IOP Conference Series: Materials Science and Engineering*, vol. 471, 2019.

[6] J. Rádl, and J. Kaiser, "Information Insufficiency Problems in Construction Projects," *Vision 2020: Sustainable Economic development, Innovation Management, and Global Growth*, vols. I-IX, pp. 5059-5070. 2017.

[7] V. Nývlt, K. Prušková, "Building Information Management as a Tool for Managing Knowledge throughout whole Building Life Cycle", *IOP Conference Series: Materials Science and Engineering*, vol. 245(4), 2017

[8] K. Prušková, "Building information management: Knowledge management principles used in the whole building lifecycle within the building information modeling technology", *18th International Multidisciplinary Scientific Geoconference, SGEM 2018*, vol. 18(2.1), pp. 209-215, 2018

[9] K. Prušková, M. Dědič, and J. Kaiser, "Possibilities of Using Modern Technologies and Creation of the Current Project Documentation Leading to the Optimal Management of the Building for Sustainable Development," *IOP Conference Series: Earth and Environmental Science*, vol. 290, 2019.

[10] P. Budský, and J. Kaiser, "Industry 4.0 Impact on Enterprise Information Systems," *Education Excellence and Innovation Management through Vision 2020*, pp. 5639-5646, 2019.

[11] J. Kaiser, "Informatics Knowledge for Setting Requirements on IFC Data and Its Suitability in Education for Chemical Engineers" *IOP Conference Series: Materials Science and Engineering*, to be published

[12] K. Prušková, "Reducing failures rate within the project documentation using Building Information Modelling, especially Level of Development", *MATEC Web of Conferences,* vol. 146, 2018.

[13] J. Myslin, and J. Kaiser, "Priority of Requirement and Method for Definition of Project Scope," *Sustainable Economic Growth, Education Excellence, and Innovation Management through Vision 2020*, vols. I-VII, pp. 4002-4013, 2017.

Fire Safety in Pre-trial Detention Facilities

A Parfyonenko [1]

[1] National Research Moscow State University of Civil Engineering, Yaroslavskoye Shosse, 26, Moscow, 129337, Russian Federation

ikbs@mgsu.ru

Abstract. This article describes the main provisions for assessing the safety of people during the evacuation from the buildings and structures in the event of a fire. The analysis of normative and scientific bases in the ensuring the safety of people during the evacuation showed the absence of psychophysiological patterns of behaviour and movement of people in the buildings of pre-trial detention facilities necessary to assess the estimated values of fire risk. The data presented in the methodology for assessing fire risks do not allow to reliably assess the probability of evacuating people in the event of a fire in such institutions, since the mode of operation of the facility is not taken into account. The key factor determining timely evacuation is the time of the beginning of the evacuation, depending on the behavioural aspects of the main functional contingent. The absence of such studies determines the need for field observations and experiments in the field of evacuation of people from the buildings of pre-trial detention facilities.

1. Introduction

The new millennium in Russia was marked by the adoption of a series of federal safety laws based on the risk assessment. The beginning was made by the adoption on Federal Law of the Russian Federation of 27.12.2002 No.184-FZ "On technical regulation", which establishes that technical regulations are adopted with a view to:

- the protection of life and health of citizens, property of individuals and legal entities, state or municipal property;
- the environmental protection, life and health of animals and plants;

"Prevention of misleading purchasers."

Realizing these goals in relation to construction projects, the Federal Law of the Russian Federation of 30.12.2009 No.384-FZ "Technical regulation on the safety of buildings and structures" puts forward (article 3) the requirements for ensuring:

1) the mechanical safety;
2) the fire safety;
3) the safety in hazardous natural processes and phenomena and (or) anthropogenic impacts; etc.

2. Methodology

Summarizing the requirements of federal safety laws, based on risk assessment, we see that the federal legislation of Russia provides for the need to ensure the safety of people in buildings and structures:

- when operating under normal conditions,

- with the threat of dangerous natural processes,
- in case of the fire.

According to the protocol No. 1, dated March 25, 2010 of the meeting of the working group of representatives of the Ministry of Regional Development and the Ministry of Emergencies of the Russian Federation to consider the draft of the Federal Law of the Russian Federation of 22.07.2008 No.123-FZ "Technical regulation on fire safety requirements" and normative documents on fire safety, the following decision was recorded:

In order to avoid duplication and contradictions in the documents establishing the fire safety requirements, it is considered appropriate to use as documents ensuring compliance with the requirements of the Federal Law of the Russian Federation of 30.12.2009 No.384-FZ "Technical regulation on the safety of buildings and structures", codes of practice in the field of fire safety.

Federal law specifies fire safety requirements [1]. Moreover, the fire safety of the buildings and structures is considered to be ensured if all the requirements of federal laws and regulations on fire safety are met. In the event that it is impossible to comply with the requirements of regulatory documents on fire safety, their justification is allowed by the calculation of the fire risks [1, article 6], where the main component is the probability of evacuation of people, the determination of which requires a reliable assessment of the safety of people. In accordance with article 53 the safety of people during evacuation is a determining factor in the formation of the space-planning structure of evacuation routes and exits in the buildings and structures [1].

The timeliness and freedom from the obstruction are two criteria for the safety of people during evacuation. The timeliness and freedom from the obstruction are expressed by the ratios:

$$t_{ev} \le t_{add} \le t_{bl} \tag{1}$$

$$D_i \le D_{add} \tag{2}$$

where:

- t_{ev} - time of evacuation of people, determined by the moment of the beginning of evacuation and the time of movement of people from their places of residence to the cross section of the evacuation route, after which the safety of evacuated people is ensured, min;
- t_{add} - the permissible time spent by people in the system (or subsystem) of evacuation routes (premises, floors, staircases, buildings) under consideration. If the permissible duration of evacuation is determined by the dynamics of dangerous factors in an emergency (for example, fire), then the time for these factors to reach levels that are dangerous to human health or life becomes necessary to complete the evacuation. Allowable time can be determined not only by dangerous, but also by technological requirements. But first of all, it is determined by the physical and psychophysiological capabilities of a person;
- t_{bl} - time of blocking escape routes by hazardous fire factors (HFF);
- D_i is the estimated value of the density of people on any part of the evacuation route;
- D_{add} - the permissible density of people on the site, determined by the required level of their comfort or safety of their stay in clusters.

To evaluate these criteria (1) and (2), software packages have been developed that allow reproducing the dynamics of the increase in HFF and the process of movement of people in buildings and structures. PyroSim or Pathfinder allow you to most adequately simulate the evacuation of people with the simultaneous spread of HFF along the evacuation routes.

Based on the provisions indicated above, it can be concluded that the fire protection system is such a system in which a person is not only an object of protection, but also a participant in his defence, as well as a participant in the protection of another person in need of assistance, since the system is forced to use his physical abilities to achieve the goals of his functioning. In this case, when a person's life is in the hands of himself or in the hands of another person (rescuer), the geometric dimensions of the path (length, width, height, slope, etc.) that will need to be overcome will play a decisive role for successful rescue or evacuation. to be in a safe place from the AFP, and the psychophysiological capabilities of people. In this regard, requirements should be imposed on evacuation routes that would ensure not only the safe conduct of evacuation, but also the salvation of the main functional contingent located in them.

According to the paragraph 4.1.3 of the Code of rules - SP (in Russian) 1.13130.2009, the protection of people on the evacuation routes is ensured by a complex of space-planning, ergonomic, structural, engineering and organizational measures. The evacuation routes within the premises must ensure the safe evacuation of people through the evacuation exits from the premises without taking into account the fire extinguishing and smoke protection used in it.

When applying expressions 1 and 2, it is also taken into account that the maximum permissible duration of the accumulation on the evacuation paths t_{acc} is 6 min (if it is exceeded, compression asphyxiation is possible) for the contingent consisting of healthy and active adults. In the flow of such people, clusters, as a rule, begin to form when the density Di reaches more than 5 people in summer clothes per m^2. The maximum intensity of movement of adults q_{max}, which is the criterion for determining the minimum required width of the evacuation routes, is in the range of the flux density D_{add} - 4-5 people/m^2.

The regularities of the relationship between the parameters of human flows were discovered as a result of studies performed by V.V. Kholshchevnikov during his dissertation for the Doctor's degree. For the first time in the history of studies of human flows, he analysed psychophysiological systems that could determine the essence of the relationships between the parameters of human flows manifested under different conditions and gave a mathematical description of the human flow as a stochastic process, and the main functional relationship V = *f* (D) in the form elementary random function. The general psychophysical law of Weber - Fechner was chosen as the main psychophysical law that describes the relationship of changes in the speed of people's movement with density, taking into account the emotional state and the type of path, as the most appropriate. The formula for determining the speed of human flow is the following:

$$\overline{V}^{e}_{Dj} = \overline{V}^{e}_{0j}\left(1 - a_j \ln \frac{D_j}{D_{0j}}\right) \tag{3}$$

where

- $\overline{V}^{e}_{0j}$ – random value of the speed of free movement (in the absence of the influence of surrounding people), depending on the type of path (j) and the level of emotional state (e) of people, m/min;
- a_j – coefficient, which is determining the degree of influence of flux density when moving along the j-th type of path;
- D_j – current value of flux density, people/m^2;
- $D_{0,j}$ – threshold value of the flux density, upon reaching which the density becomes a factor affecting the speed of movement, people/m^2.

The established pattern formed the basis for normalizing evacuation routes and exits in the event of a fire. Thus, the fire protection system is such a system in which a person is not only an object of protection, but a participant in his defense (self-defense), since the system is forced to use its natural capabilities (without evacuating the fire risk value, it is 2-3 orders o order lower than the standard value) to achieve the goals of its functioning.

3. Results and discussions

The methodology makes it possible to assess fire risks for all buildings of various classes of functional fire hazard [2]. But the applicability of this methodology to institutions providing isolation of suspects, accused, defendants and convicts (pre-trial detention facilities) is doubtful. Firstly, similar institutions according to clause 22.6 of the Code of rules - SP (in Russian) 247.1325800.2016 “Detention facilities of the penitentiary system. Design Rules” refers to the class of functional fire hazard - F1.2, for which the methodology defines the time for the start of evacuation in the range from 2 to 3 minutes, depending on the type of the systems of warning and evacuation control. Class F1.2 includes the sleeping buildings of sanatoriums and rest houses of a general type of motels, boarding houses, campsites, hotels and hostels.

4. Conclusions

Accordingly, the use of the data presented in the methodology does not correspond to the real mode of operation of the facility. The time for the start of evacuation in such institutions primarily depends on the psychological characteristics of the person and the opening time of the cells for detainees, which leads to a significant increase in the time spent on the start of the evacuation. For example, a fire in 2015 in a pre-trial detention centre No. 1 in Ulyanovsk claimed the lives of four people, with eight people injured. The cause of the fire was a short circuit in the electrical wiring, the cause of death was poisoning with toxic combustion products as a result of untimely people entering the safe zone, where the time for the start of the evacuation became a key factor. Secondly, in the methodology there are no psychophysiological regularities of the movement of human flows in the pre-trial detention facilities buildings. The use of motion parameters of adult healthy people will lead to an unreasonable assessment of the fire safety of such buildings in connection with the heterogeneity of the composition of human flows in the pre-trial detention facilities buildings.

Studies on the study of the psychophysical laws of behaviour and movement of people in case of fire in the pre-trial detention facilities buildings were not found in both domestic and foreign literature [3-17]. The study of this issue is very relevant, since without taking into account the above aspects, it is impossible to correctly assess the calculated values of a firefighter during the design, construction and operation of pre-trial detention facilities.

References

[1] Federal Law of the Russian Federation of 22.07.2008 No.123-FZ "Technical regulation on fire safety requirements" (in Russian), 2008.
[2] The methodology for determining the calculated values of fire risk in buildings, structures and structures of various classes of functional fire hazard. Appendix to the Order of the Ministry of Emergencies of Russia dated June 30, 2009 No. 382, subject to the changes approved by Order No. 632 of the Ministry of Emergencies of the Russian Federation of December 2, 2015 (in Russian), 2015.
[3] A. Milinsky, *Study of the evacuation of buildings for public use*, Dis. Ph.D. 05.23.10 Moscow (in Russian), 1951.
[4] V. Predtechenskiy, *Human Streams in Buildings*, Moscow: Stroyizdat (in Russian), 1982.
[5] V. Kalintsev V, *Designing cinema halls taking into account the movement of human flows*, Dis. Ph.D. 05.23.10 Moscow (in Russian), 1966.
[6] R. Duvidzon, *Design of sports facilities taking into account the movement of human flows,* Dis.

Ph.D. 05.23.10 Moscow (in Russian), 1968.
[7] V. Kholshchevnikov, *Optimization of ways of movement of human flows. High-rise buildings*, Dis. Ph.D. 05.23.10 Moscow (in Russian), 1969.
[8] R. Grigoryants, *The study of long-existing human flows*, Dis. Ph.D. 05.23.10 Moscow (in Russian), 1973.
[9] Y. Alekseev, *Formation and movement of human flows in the aisles of spectacular structures*, Dis. Ph.D. 05.23.10 Moscow (in Russian), 1978.
[10] V. Gvozdyakov, *Patterns of movement of human flows in transport and communication facilities*, Dis. Ph.D. 05.23.10 Moscow (in Russian), 1978.
[11] M. Eremchenko, *The movement of human flows in school buildings*, Dis. Ph.D. 05.23.10 Moscow (in Russian), 1978.
[12] V. Kopylov, *Research of parameters of the movement of people during forced evacuation*, Dis. Ph.D. 05.23.10 Moscow (in Russian), 1974.
[13] A. Parfyonenko, *Rationing of fire safety requirements for evacuation routes and exits in the buildings of preschool educational institutions*, Dis. Ph.D. (scientific adviser Kholshchevnikov V) Moscow AGPS of the Ministry of Emergencies of the Russian Federation (in Russian), 2012.
[14] V. Kholshchevnikov, *The study of human flows and the methodology for standardizing the evacuation of people from buildings in case of fire*, Moscow MIPB of the Ministry of Internal Affairs of the Russian Federation (in Russian), 1999.
[15] O. Aydn, "A Qualitative Approach to Children of Developing Countries from Human Behavior Point of View," *Proceedings of the 2nd International Symposium on Human Behaviour in Fire*, pp. 531 – 538, 2011.
[16] D. Bruck, "Community based research on the effectiveness of the home smoke alarm in waking up children," *Proceedings of the 4th International Symposium on Human Behaviour in Fire*, pp. 335 – 344, 2009.
[17] A. Larusdottir, and A. Dererichs,"Evacuation dynamics of children – walking speed, flow through doors in day care centers," *Proceedings of the 5th International Conference on Pedestrian and Evacuation Dynamics*, pp. 67 – 79, 2010.

Aging Simulation of Plastic Products due to Weather Conditions

Petra Machova [1], Lucie Crhakova [1], Michal Vejvoda [1], Michal Kraus [1]

[1] Institute of Technology and Business in České Budějovice, Department of Civil Engineering, 370 01, České Budějovice, Czech Republic

petra.machova@mail.vstecb.cz

Abstract. The paper describes the aging of plastic products due to weather conditions. It focuses on the analysis of aging and weathering of plastic products, or parts of building structures. As part of the research, the Xenotest Alfa device was used to simulate the weathering of plastic products. Xenotest Alfa simulates, for example, variable temperature, solar radiation, humidity and different types of climate. Samples of plastic and wood-plastic products were placed in the device, where the test of accelerated degradation and light stability was performed. The motivation for using the device is the effect of UV radiation on the samples, which causes the aging of the material and its gradual degradation, which may in the future cause the product to malfunction or degrade its color, as shown by this research.

1. Introduction

Polymers are substances that simultaneously extend their action through various industries. The reason is the replacement of common materials, such as metal, wood or glass. Polymers are chemically pure macromolecular synthetic or natural substances that are made up of large chains of large molecules [1,2]. The basic building block of polymers is a carbon atom. Macromolecular chains can be formed by different or the same building blocks that are connected by a covalent bond. In addition to the carbon atom, these building blocks can also be hydrogen atoms and, to a lesser extent, oxygen, nitrogen and chlorine atoms.

Polymer plastics offer a wide range of properties such as lightness, durability, color fastness, transparency and ease of processing [3]. Due to the growing use of polymeric materials, initiation is being formed to understand the degradation processes of these substances. The process of polymer degradation is explained as a change in the structure and properties of polymers caused by decomposition reactions. Degradation of the material occurs already during the processing phase, when polymeric substances are exposed to high temperatures, pressures and mechanical stress [4]. During the processing of the molten polymer, for example, the chains are degraded and cleaved, and flue gases are released. Then, when the material is processed or in the solid state, a second degradation step takes place. This degradation is less intense but long-lasting. It is caused mainly by weather influences. The changes are usually manifested by loss of gloss, discoloration, loss of transparency or cracks on the surface [5].

All materials are exposed to long-term exposure to the outdoor environment - specifically to weather conditions, which with their variability create a diverse range of operating conditions. Specifically, weather conditions, which with their variability form a diverse range of operating conditions. These conditions result in lower service life of the material, loss of the original mechanical or physical properties. In order to predict the life and stabilization of a given material, it is first necessary to understand the principles and mechanisms of degradation. The ideal solution would be to verify the polymer properties in real-time, which is often very difficult or even impossible. High demands are placed on materials, mainly for their service life, in some cases up to decades. Given the required operating time and the rapid development of new materials, it is difficult to wait for such a long period of time to verify the properties of these substances. For life prediction, they were thus invented methods that are called methods of accelerated weathering.

2. Materials and methods

The long-term effect of the external environment on the material causes changes that can be described as the weathering of polymers. This process is mainly influenced by light (or sunlight), UV radiation, temperature and its changes, humidity, oxygen and ozone. The sum of these external factors affecting the material is considered to be climatic conditions. There are other factors such as chemicals, the action of microorganisms, dust, air flow or other influences, but they are monitored only in special cases.

The greatest influence on the aging (weathering) of polymers has solar radiation, specifically ultraviolet (UV) radiation, which is part of sunlight. Ultraviolet rays usually break the bond between two atoms in the chain of a macromolecule, and the macromolecule breaks down into smaller units that react easily with atmospheric oxygen. We call this process photooxidation.

With the growing share of plastics in the outdoor environment, especially in the automotive industry, it is necessary to perform weather aging tests. It is with regard to the weather that the aging of polymers can be observed in either a natural or artificial way. In the method of accelerated aging, the main benefit is the fact that the results can be observed in a significantly shorter period of time. However, these results are only close to real natural aging, so they are only approximate. In natural weathering, however, it is necessary to wait a long period of time, which does not correspond to the speed of material development. Evaluation of natural and artificial weathering of polymers is evaluated according to the following Table 1.

Table 1. Evaluation of natural and artificial weathering of polymers.

External appearance of the exposed part)	Loss of gloss - discoloration - formation of cracks, depressions or irregularities - erosion by water, dust - cracking of the test specimen
Mass (weight)	Weight change - loss of plasticizers
Mechanical properties	Impact toughness, elongated at break, tensile strength, flexural strength
Dielectric properties	Surface insulation resistance, volume electrical resistance electrical breakdown strength
Mechanical properties of the exposed surface	Brittleness of the surface layer, hardness of the surface layer

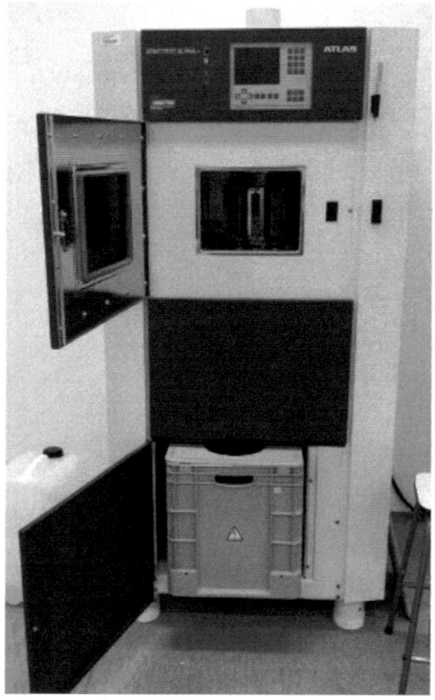

Figure 1. Xenotest Alfa weathering (aging) simulator.

Devices that can create conditions simulating natural weather are so-called veterometers. These devices can be used to simulate, for example, variable temperature, solar radiation, humidity, or various types of climate. The basic criterion according to which the residence time of bodies in the instrument is determined is the equivalent of the amount of light energy incident on an area unit of samples.

As part of the research, the aging of plastic products was simulated and analyzed using the Xenotest Alfa for lightfastness tests and simulation of weather conditions. All samples of plastic and wood-plastic products were placed on holders rotating on a carousel around a xenon lamp, and then an accelerated degradation and light stability test was performed.

The measurement was performed according to the standard CSN EN ISO 4892-2 from 2013 (64 0152). The measurement lasted a total of 1,000 hours. Exposure settings for filters using daylight simulating filters (artificial aging) are illustrated in table 2.

Table 2. Default Xenotest Alfa settings for the experiment.

Cycle no.	Exposure time (*minutes*)	Irradiation intensity (*E, W/m²*)	Black standard temperature (*BST, °C*)	Chamber temperature (*CHT, °C*)	Relative humidity (*RH, %*)
1	102.00 (drying)	60.00	65.00	38.00	50.00
	18.00 (water spray)	60.00	OFF	38.00	OFF

3. Results and discussions

The research was simulated and analyzed the aging of plastic products using Xenotest Alfa for lightfastness testing and simulation of weather conditions. A total of 17 samples were placed in Xenotest Alfa. The list of samples of tested materials is in table 3. The evaluation of the artificial weathering of the tested polymers is given in table 4.

Table 3. Overview of tested materials.

No.	Manufacturer - product	Material	Application	Lifetime [a]
1	Lanit plast - Marlon ST IR Blue	Cellular polycarbonate	Roofing and vertical glazing	10-year
2	Woodplastic – Style plus Inox	Wood plastic	Terraces	25-year
3	Rehau - windowsill	PVC	Windowsill	-
4	Inoutic - Adorn plastic window white	PVC	Window frame	-
5	WPC – Legacy Moca	Wood plastic with foil	Terrace, railings, fence	30-year color
6	WPC – Premium forest Pallisander	Wood plastic	Terraces	25-year color
7	WPC – Prémium star Inox	Wood plastic	Terraces	25-year color
8	MAX WPC - terrace board dark brown	Wood plastic	Terraces	25-year color
9	Polycasa Příbram - Elventa polycarbonate	Polycarbonate with UV filter	Translucent strips, roofing	15-year
10	MAX WPC - Lock tile gray	Wood plastic	Terraces	25-year color
11	Lanit plast – Marlon ST Bronze	Cellular polycarbonate	Roofing and vertical glazing	10-year
12	Inoutic - Adorn plastic window black	PVC	Window frame	-
13	Polycasa - Elventa corrugated PVC, profile 76/18	PVC with UV filters	Roofing	-
14	Ecolux – Societe Ondex Cristal Jaune	PVC with UV filters	Roofing	5-year
15	Irpen Group	Polycarbonate	Roofing	-
16	MAX WPC - Anthracite interlocking tiles	Wood plastic	Terraces	25-year
17	Freelite – Corrugated Translucent PVC Sheet Greca	PVC	Roofing	-

[a] Warranty stated by the manufacturer.

Table 4. Evaluation of tested materials.

No.	External appearance of the exposed part			Weight deviation (*g*)	
	Loss of shine	Change of color	Surface damage	Pre-weight	After-weight
1	Yes	Minor	Surface damage	4.12	4.12
2	-	Major	No	63.82	63.14
3	Yes	Medium	No	3.40	3.39
4	No	Minor	No	12.98	12.97
5	No	Minor	No	134.58	134.58
6	Yes	Major	Surface damage	113.55	113.53
7	-	Major	No	47.54	47.53
8	Yes	Major	Significant surface damage	39.2	39.00
9	No	No	No	9.32	9.25
10	-	Major	No	33.83	33.60
11	Yes	Minor	Surface damage	2.41	2.40
12	No	Minor	No	15.12	15.12
13	Yes, Minor	Minor	No	2.79	2.79
14	No	No	No	2.69	2.69
15	No	No	No	12.19	12.19
16	Yes	Medium	No	35.63	35.63
17	Yes	Major	No	3.55	3.55

The appearance of the samples before and after exposure to the weathering simulation can be seen below. The sample is tested on the left after testing and on the right before testing.

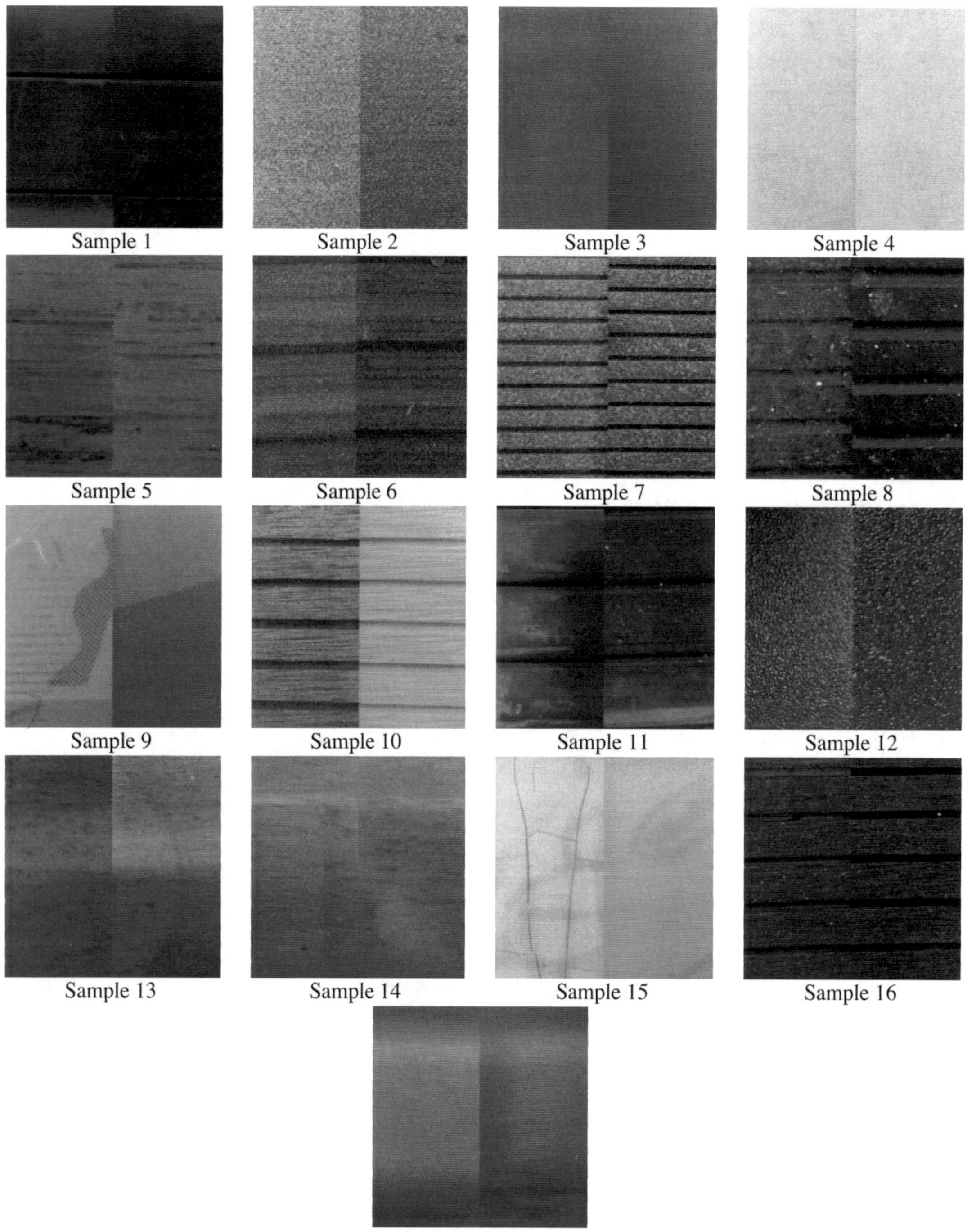

Figure 2. Appearance of samples before and after exposure to weather aging simulation.

4. Conclusions

The paper describes the process of polymer degradation. Samples of plastic and wood-plastic products were placed in the device Xenotest Alfa, which can create conditions simulating weather conditions. Thanks to this simulation, the aging of polymers can be observed rapidly. Before and after the simulation, the external appearance of the exposed part and the change in weight are monitored. Evaluation of all tested materials is in the Results chapter. The most significant changes occurred in cellular polycarbonates, polyvinyl chloride (PVC) without UV filters and especially in wood plastics. The manufacturers of the tested wood-plastics guarantee a color fastness of 25 years, while all of them have significantly changed their color and one of them has even significantly damaged the surface. This study contributes information on the aging of plastics and wood plastics. This knowledge is necessary and useful for the development of new materials.

Acknowledgment(s)
The authors gratefully acknowledge the financial support received from the Institute of Technology and Business, under project no SVV202006 Simulation of aging of plastic products due to weather conditions.

References

[1] S. E. Gad, "*Polymers*," Encyclopedia of Toxicology (Third Edition), pp. 1045-1050, 2014.
[2] S. Edmonson, and M. Gilbert, "*Chapter 2 - The Chemical Nature of Plastics Polymerzation*," Brydson's Plastics Materials, Eight edition, pp. 19-37, 2017.
[3] A. Shrivastava, "*3-Plastic Properties and Testing*," Introduction to Plastic Engineering, Plastic Design Library, pp. 49-11, 2018.
[4] J. Qin, J. Jiang, Y. Tao, S. Zhao, W. Zeng, Y. Shi, T. Lu, L. Guo, S. Wang, X. Zhang, G. Jie, J. Wang, and M. Xiao, "Sunlight tracking and concentrating accelerated weathering test applied in weatherability evaluation and service life prediction of polymeric materials: A review," *Polymer Testing*, 106940, 2020.
[5] B. Singh, and N. Sharma, "Mechanistic implications of plastic degradation," *Polymer Degradation and Stability*, vol. 93, Issue 2008, pp. 561-583, 2008.

AUTHOR INDEX

Institute of Physics
Dirac House, Temple Back
Bristol BS1 6BE UK

ISSN: 1757-8981
ISBN 978-1-7138-3342-0

CURRAN ASSOCIATES INC.
proceedings.com